AF809054

PREFACE

1. Scope

This publication describes multi-Service tactics, techniques, and procedures (MTTP) for consideration and use during the tactical employment of nonlethal weapons (NLW) in support of warfighting personnel conducting training and tactical operations. This publication—

a. Provides an overview of NLW and its relationship to deadly force.

b. Provides NLW system description.

c. Describes the capability requirements of NLW.

d. Discusses fundamental concepts and training requirements involved with NLW.

e. Discusses the tactical employment considerations of NLW.

f. Discusses lessons learned from previous NLW use.

2. Purpose

a. This publication provides a single-source, consolidated reference on the tactical employment of NLW and supporting systems.

b. This publication is not intended to restrict the authority of the commander from organizing the force and executing the mission in a manner deemed appropriate.

3. Application

a. This publication provides commanders and their staffs unclassified guidance for NLW employment and planning. Commanders and staffs can use this publication to aid in the tactical employment of NLW during exercises and contingencies.

b. The United States (U.S.) Army, Marine Corps, Navy, Air Force, and Coast Guard approved this multi-Service publication for use.

4. Implementation Plan

Participating service command offices of primary responsibility will review this publication, validate the information and reference, and incorporate it in service manuals, regulations, and curricula follows.

Army. The Army will incorporate the procedures in this publication in U.S. Army training and doctrinal publications as directed by the Commander, U.S. Army Training and Doctrine Command (TRADOC). Distribution of this publication is in accordance with (IAW) initial distribution number (IDN) 115757.

Marine Corps. The Marine Corps will incorporate the procedures in this publication in U.S. Marine Corps doctrinal and training publications as directed by the Commanding General, U.S. Marine Corps Combat Development Command (MCCDC). Distribution of this publication is IAW Marine Corps Doctrinal Publication System (MCPDS).

Navy. The Navy will incorporate the procedures in this publication in U.S. Navy doctrinal and training publications as directed by the Commander, Navy Warfare Development Command (NWDC). Distribution of this publication is IAW Military Standard Requisition-

MARINE CORPS PCN: 143 000057 00
Air Force Distribution: F

ing and Issue Procedures (MILSTRIP) Desk Guide and Navy Standing Operating Procedure (NAVSOP) Pub 409.

Air Force. The Air Force will validate and incorporate appropriate procedures IAW applicable governing procedures. Distribution is IAW Air Force Instruction (AFI) 33-360.

Coast Guard. The Coast Guard will incorporate the procedures in this publication in U.S. Coast Guard doctrinal and training publications as directed by the Assistant Commandant for Operations, Director of Operations Policy, U.S. Coast Guard (USCG). Distribution of this publication is IAW Coast Guard standing operating procedures.

5. User Information

a. TRADOC, MCCDC, NWDC, Air Force Doctrine Center (AFDC), USCG, and the Air Land Sea Application (ALSA) Center developed this publication with the joint participation of the approving service commands. ALSA will review and update this publication as necessary.

b. This publication reflects current joint and service doctrine, command and control organizations, facilities, personnel, responsibilities, and procedures regarding NLW employment. Changes in service protocol, appropriately reflected in joint and service publications, will likewise be incorporated in revisions of this document.

c. We encourage recommended changes for improving this publication. Key comments to the specific page and paragraph and provide rationale for each recommendation directly to—

***FM 3-22.40** (FM 90-40)
***MCWP 3-15.8**
***NTTP 3-07.3.2**
AFTTP(I) 3-2.45
***USCG PUB 3-07.31**

FM 3-22.40 (FM 90-40) — **U.S. Army Training and Doctrine Command, Fort Monroe, Virginia**

MCWP 3-15.8 — **U.S. Marine Corps Combat Development Command, Quantico, Virginia**

NTTP 3-07.3.2 — **U.S. Navy Warfare Development Command, Newport, Rhode Island**

AFTTP(I) 3-2.45 — **Headquarters, Air Force Doctrine Center, Maxwell Air Force Base, Alabama**

USCG Pub 3-07.31 — **U.S. Coast Guard, Washington, D.C.**

15 January 2003

NLW
Tactical Employment of Nonlethal Weapons

TABLE OF CONTENTS

Page

*This publication supersedes FM 90-40, MCRP 3-15.8, NWP 3-07.31, USCG Pub 3-07.31, dated 6 October 1998.

FIGURES

Tables

EXECUTIVE SUMMARY

Tactical Employment of Nonlethal Weapons (NLW)

Today's world events mandate a need to project nonlethal force across all levels of war to enable our warfighters and leaders to effectively deal with a host of traditional as well as nontraditional threats. Now more than ever, the minimal level of tolerance for collateral damage and loss of human life, coupled with the tendency for the typical adversary to exploit the rules of engagement (ROE) to his benefit, necessitates an effective and flexible application of force through nonlethal weapons. Much like a rheostat switch in which power can be dialed up or down as desired, NLW provide tools to allow a commander to employ sufficient force to accomplish an objective without requiring the destruction of an enemy or the habitat. The intent of employing NLW is not to add another step in the progression of escalation with an adversary, but to add another tool to use anywhere along that continuum. This publication provides initial guidance for employing NLW in a tactical environment.

Overview of NLW

Chapter I describes the background, definition, and the concept description of NLW. This chapter discusses the development and employment of NLW, and explains the relationship of NLW to deadly force capabilities. The final portion of the chapter reviews policy per Office of Secretary of Defense instructions on employing NLW.

Purpose, System Description, and Organization

Chapter II describes the purpose of NLW and gives a general description of currently available commercial or government off-the-shelf equipment and munitions. The chapter also addresses organization structure, to include system location/distribution, personnel requirements, and mission requirements/tactical structure.

New Capability Requirements

Chapter III describes the applicability of NLW across the range of military operations. Additionally, this chapter discusses the core capabilities as stated in the Joint Mission Area Analysis (JMAA) as described in the Department of Defense Joint Concept for NLW.

Training With the NLW Capability Set

Chapter IV describes training requirements, topics, methods, locations, and a program of instruction from the Interservice Nonlethal Individual Weapons Instructor Course currently available for NLW options. This chapter also addresses additional training available from civilian law enforcement agencies.

Employment of NLW Capabilities

Chapter V establishes the fundamental understanding of the force continuum needed to employ NLW. It outlines employment prerequisites, mission-planning factors, and personnel requirements needed to employ NLW options. It further delineates NLW tasks according to the JMAA and provides considerations for preparation for unit tactical operations. This chapter also discusses employment objectives for NLW and ROE development. This chapter ends with multiple examples of service-specific crowd control formations used by units equipped with NLW.

Lessons Learned

Chapter VI focuses on NLW lessons learned from recent military operations. The chapter discusses how nonlethal technology has changed the way forces are looking at their adversaries in using NLW. It also covers the use of NLW along the force continuum in conjunction with ROE and the media.

PROGRAM PARTICIPANTS

The following commands and agencies participated in developing this publication:

Joint

U.S. Joint Forces Command, Suffolk, VA
U.S. European Command, J3/J5, APO AE
U.S. Pacific Command, J3832, Camp H. M. Smith, HI
U.S. Central Command, J3, MacDill AFB, FL
U.S. Southern Command, SCJ5-PS, Miami, FL
U.S. Space Command, SPJ5X, Peterson AFB, CO
U.S. Special Operations Command, HQ USSOCOM, SOOP-PTE, MacDill AFB, FL
Joint Nonlethal Weapons Directorate, Quantico, VA
Interservice Nonlethal Individual Weapons Instructor Course, Fort Leonard Wood, MO

Army

Headquarters, Department of the Army, ODCSOPS (DAMO-FDT), Pentagon, Washington, D.C.
U.S. Army Training and Doctrine Command, DCSDCS (ATDO-A), Fort Monroe, VA
U.S. Army Training and Doctrine Command, DCSDEV (ATCD-SN), Fort Monroe, VA
U.S. Army Infantry Center, DCD, Fort Benning, GA
U.S. Army Military Police School, ATTN: DOTD, MANSCEN (ATZT-DT-WM-M), Fort Leonard Wood, MO
USAJFKSWCS, DOTD, Fort Bragg, NC
National Ground Intelligence Center, Charlottesville, VA

Marine Corps

Marine Corps Combat Development Command, Quantico, VA
MARCORSYSCOM, Quantico, VA
USMC Warfighting Lab, Quantico, VA
MCSFT Company, MCSF Battalion, Chesapeake, VA
2d FAST MCSF Battalion, Yorktown, VA
1st MARDIV Military Police Company, HQ Battalion, Camp Pendleton, CA
Marine Corps Detachment Military Police and Chemical School, Fort Leonard Wood, MO
Special Operations Training Group, I Marine Expeditionary Force, Camp Pendleton, CA
Special Operations Training Group, II Marine Expeditionary Force, Camp Lejeune, NC

Navy

Commander, Navy Warfare Development Command (NWDC), Newport, RI
USN Surface Warfare Center, Crane, IN
USN Surface Warfare Center, Dahlgren, VA

Air Force

AFDC/DJ, Langley AFB, VA
Headquarters, ACC/SFOT, Langley AFB, VA
Air Force Security Forces Center/SFOR, Lackland, AFB, TX
Headquarters, USAF/XOF, Washington, D.C.

Coast Guard

Commandant, U.S. Coast Guard, (G-OPL-2), Washington, D.C.

Chapter I

Overview of Nonlethal Weapons

1. Background

a. In today's operational environment, United States (U.S.) forces regularly perform missions that were almost inconceivable a few decades ago. In this new environment, lethal firepower or the threat of its use may no longer be the default solution to all crises or problems. Senior leaders face a new level of public and media sensitivity and scrutiny concerning the proper role of the military as an instrument of national power. Field commanders must understand these sensitivities and attempt to achieve an appropriate military force. Junior leaders must apply the resulting decisions wisely, often in changing situations filled with uncertainty and danger.

b. Throughout history, changes in culture and technology have influenced the character of military force and the manner in which it is employed. These changes are an attempt to maximize military force in the force continuum. In the complex and changing political and social domain of the modern world, nonlethal capabilities may offer the opportunity to increase or maximize the military as an instrument of national power.

c. Nonlethal weapons (NLW) must now be considered an integral part of joint operational capabilities. Joint forces anticipate that NLW technology will advance and provide additional options for force commanders.

2. Definition

a. Department of Defense Directive (DODD) 3000.3, Policy for Nonlethal Weapons, 9 July 1996, defines NLW as "weapons that are explicitly designed and primarily employed so as to incapacitate personnel or materiel while minimizing fatalities, permanent injury to personnel, and undesired damage to property and the environment." Furthermore, "unlike conventional lethal weapons that destroy their targets principally through blast, penetration, and fragmentation, nonlethal weapons employ means other than gross physical destruction to prevent the target from functioning. Nonlethal weapons are intended to have relatively reversible effects on personnel and materiel."

b. The term "nonlethal" does not mean zero mortality or nonpermanent damage; these are goals and not guarantees of these weapons. NLW can add flexibility to combat operations and enhance force protection by providing an environment in which friendly troops can engage threatening targets with reduced risk of noncombatant casualties and collateral damage.

3. Concept Description

Nonlethal capabilities within the context of the service capability sets (see Appendix E) may provide armed forces additional offensive and defensive tools to assist in the incapacitation and destruction of an enemy, capture of important intelligence sources, crowd control, civil disturbances, enforcing mandates, and force protection in accordance with the specified rules of engagement (ROE) for a specific operation. These operations include small-scale contingencies (SSC), military operations other than war (MOOTW), stability and support operations (SASO), and civil disturbances. The U.S. Army's Task Force (TF) FALCON in Kosovo and U.S. armed forces located at Camp X-RAY in Guantanamo Bay, Cuba are superb examples of the use of NLW to prevent escalation within an operation.

4. Background on Development and Employment

a. Military forces have long used nonlethal force to influence behavior of people and nations, defeat adversaries with minimum use of lethal capabilities, and weaken adversaries to defeat them with conventional arms. Examples of nonlethal means include a show of force, physical obstacles, riot control agents (RCA), noise to create or enhance psychological effects, smoke and obscurants to mask operations or defeat homing and guidance mechanisms, and using light to disorient combatants. These nonlethal means will remain relevant in future operations and are a part of an evolutionary process of weapons development. Joint operations in Somalia provided increased awareness of the need to incorporate nonlethal options into the fighting capabilities. Therefore, the Department of Defense (DOD) established the Joint Nonlethal Weapons Program in 1996 to research, develop, and acquire these capabilities.

b. Increased interaction between U.S. forces and civilian populations has become a feature of the contemporary operational landscape that is likely to remain the case in the future. Three factors account for this forecast:

(1) Worldwide patterns of population growth and migration have resulted in an expansion of urban culture within established industrialized nations, as well as many pre-industrial and emerging third world societies (60 percent of the world's population live in built-up areas). The prevalence of urbanization in many crisis-prone regions of the world creates the potential for large, vulnerable groups of noncombatants to be entrapped in volatile confrontations involving the deployment of military forces in military operations in urban terrain (MOUT).

(2) U.S. forces increasingly operate in the challenging environment known as MOOTW; a category that includes such missions as humanitarian assistance, disaster relief, noncombatant evacuations, and the various types of peace operations. These operations commonly involve close and continual interaction between U.S. forces and noncombatant civilians. Some MOOTW scenarios include the presence of paramilitary forces, armed factions, or rogue elements that present a constant, but uncertain threat to joint forces. In these situations, the mission of military forces includes preventive tasks. In other words, U.S. forces enhance mission accomplishment by preventing individuals or factions from carrying out specified undesirable activities, such as rioting, looting, attacking, harassing, or threatening. Often, the adversary blends in with the local population of innocent citizens. At other times, sectors of the local population may rise against U.S. forces and become active participants in acts of violence. Factional alignments, the level of violence, and the threat to mission accomplishment, may change frequently with little or no warning. Under such circumstances, the identity of adversaries is uncertain and the use of deadly force for purposes other than self-defense may be constrained by ROE or the judgment of the on-scene commander.

(3) From a defense-in-depth or situational understanding perspective, the force protection, homeland security, and the antiterrorism missions, require that forces assess the use of NLW as a response to potential asymmetric attacks. Asymmetric threats run the spectrum of adversarial use of innovative or nontraditional strategies, tactics, or technologies, to exploit vulnerabilities and strike a decisive blow while avoiding the strengths of an opponent. The objectives of these strikes are intended to undermine national will. Terrorist attacks such as those directed against the USS Cole, World Trade Center, and Pentagon, highlight the severity of the asymmetric threat and its resultant effects.

5. Relationship of NLW to Deadly Force

a. NLW provide "the lesser means" that could be considered prior to employing deadly force. The services definition of deadly force is as follows:

(1) U.S. Army – "Force that a person uses causing, or that a person knows, or should know, will create a substantial risk of causing death or serious bodily harm." (Army Regulation (AR) 190–14, Military Police Carrying of Firearms and Use of Force for Law Enforcement and Security Duties, 12 March 1993).

(2) USMC – "That force which a person uses with the purpose of causing death or serious bodily harm or which a reasonable and prudent person would consider likely to create substantial risk of causing death or serious bodily harm. It will be used as a last resort when all lesser means have failed." Marine Corps Order (MCO) 5500.6F, Arming of Security and Law Enforcement Personnel and the Use of Force, 20 July 1995.

(3) USN – "Force that a person uses causing, or that a person knows or should know would create a substantial risk of causing death or serious bodily harm." Secretary of The Navy Instruction (SECNAVINST) 5500.29B, Use of Deadly Force and the Carrying of Firearms by Personnel of the Department of the Navy in Conjunction with Law Enforcement, Security Duties and Personal Protection, 29 September 1992.

(4) USAF – "Force that causes, or that you know has a substantial risk of causing death or serious bodily harm." Air Force Instruction (AFI) 31-207, Arming and Use of Force by Air Force Personnel, 1 September 1999.

(5) USCG – "Deadly force is any force that is likely to cause death or other serious physical injury." (COMDTINST M16247.1B, Maritime Law Enforcement Manual, 9 August 2001.)

b. The commitment of military power to resolve crises has traditionally involved the use of deadly force or the implicit or explicit threat of the use of deadly force. Military units are trained, organized, and equipped for this purpose. A force armed only with traditional military weapons normally has two options for enforcing compliance: threating deadly force and applying deadly force. This limitation creates a critical vulnerability that belligerents may quickly discern and use to their advantage.

c. Nonlethal capabilities provide a wider range of options that augment, but do not replace traditional means of deadly force. The option to resort to deadly force must always remain available when the commander believes it is appropriate to the mission. DODD 3000.3, states "the availability of NLW shall not limit a commander's inherent authority and obligation to use all necessary means available and to take all appropriate action in self-defense." The existence of nonlethal capabilities does not represent the potential for "nonlethal war," and unrealistic expectations to that effect must be vigorously avoided. Noncombatant casualties, to include serious injuries and fatalities, are often an unavoidable outcome when employing military power, regardless of NLW availability. Augmenting deadly force with NLW should be fundamental to planning and executing any operation where the commander is contemplating deadly force.

d. The commander must clearly articulate the roles of NLW within the ROE. NLW is an additional means of employing force throughout the force continuum for the particular purpose of reducing the probability of death or serious injury to noncombatants or belligerents. Using deadly force must always remain an inherent right of individuals in instances when they, their fellow servicemen, or personnel in their charge are threatened with death or serious bodily harm. NLW add flexibility to combat operations and enhance force protection by providing the field commander with options to engage threatening targets with limited risk of noncombatant casualties and collateral damage.

e. Commanders and public affairs officers (PAO) must prepare personnel to address media questions and concerns regarding the role of NLW. Operational experience indicates that new capabilities give rise to significant media interest. Personnel participating in interviews or briefings must be prepared to address the role of NLW. It must be clear that the presence of NLW in no way indicates abandoning the option to employ deadly force in appropriate circumstances.

6. Advantages of Employing NLW

NLW provide commanders the flexibility to influence the situation favorably with increased safety to friendly forces and reduced risk of both noncombatant fatalities and collateral damage. Some advantages follow:

a. NLW can be more humane and consistent with the political and social implications implicit in humanitarian and peacekeeping missions.

b. The force that properly employs nonlethal options can gain advantages over those who rely on lethal options alone. Because the degree of provocation required to employ NLW is substantially less than lethal options, the force can take a more proactive position that may diminish the likelihood of having a situation escalate to a point where deadly force is required to resolve a conflict.

c. NLW options can be less likely to provoke others. However, commanders should be cautioned using NLW does not preclude an undesired result.

d. Demonstrated restraint can greatly diminish feelings of anger and remorse when deadly force is required after nonlethal options fail.

e. Using NLW can facilitate post incident stabilization by reducing populace alienation and collateral damage.

f. NLW can reduce the possibility of injury to friendly forces compared to forces without NLW.

7. Policy Per Office of the Secretary of Defense Instructions on Employment of NLW

a. DODD 3000.3 states, "nonlethal weapons, doctrine, and concepts of operation shall be designed to reinforce deterrence and expand the range of options available to commanders. Nonlethal weapons should enhance the capability of U.S. Forces to accomplish the following objectives:

(1) Discourage, delay, or prevent hostile actions.

(2) Limit escalation.

(3) Take military action in situations where use of lethal force is not the preferred option.

(4) Better protect our forces.

(5) Temporarily disable equipment, facilities, and personnel.

(6) Nonlethal weapons should also be designed to help decrease the post-conflict costs of reconstruction."

b. DODD 3000.3 continues by stating, "The availability of nonlethal weapons shall not limit a commander's inherent authority and obligation to use all necessary means available and to take all appropriate action in self-defense. Neither the presence nor the potential effect of nonlethal weapons shall constitute an obligation for their employment or a higher standard for employment of force than provided for by applicable law. In all cases, the United States retains the option for immediate use of lethal weapons, when appropriate, consistent with international law."

c. DODD 3000.3 further states, "Nonlethal weapons shall not be required to have a zero probability of producing fatalities or permanent injuries. However, while complete avoidance of these effects is not guaranteed or expected, when properly employed, nonlethal weapons should significantly reduce them as compared with physically destroying the same target." Nonlethal weapons may be used in conjunction with lethal weapon systems to enhance the latter's effectiveness and efficiency in military operations. This shall apply across the range of military operations to include those situations where overwhelming force is employed."

d. According to paragraph 3.21 of the Interium Guidance issued by the Deputy Secretary of Defense, 30 October 2002, all developmental weapon systems must be submitted for legal review and must receive favorable findings before they can be incorporated into the force. This review includes examination for consistency with applicable international and domestic laws, including the laws of war, various arms control treaties, and protocols. U.S. forces can be assured that any NLW they are issued have passed this critical test. Medical and policy reviews should also be considered prior to fielding any new weapon systems. Commanders and operators should consult their specific ROE for guidance.

e. Some capabilities, although designed to minimize fatalities and serious injuries, may have effects that could actually discourage their use. Therefore, the basic decision to employ military force in defense of national interests is usually a matter of intense public concern. The manner in which that force is exercised is subjected to the same scrutiny.

f. The presence of NLW on the battlefield does not mean they must be used. It also does not mean they must be used prior to the use of lethal force. In all cases, the U.S. retains the option for immediate use of lethal weapons when the on-scene commander deems it appropriate. Additionally, the immediate use of lethal weapons must be consistent with U.S. and international law and the ROE.

Chapter II

PURPOSE, SYSTEM DESCRIPTION, AND ORGANIZATION

1. Purpose

The purpose of NLW is to enhance the ability of military commanders to conduct operations across the spectrum of conflict. Over the past several years, military operations have gravitated towards the lower end of this spectrum and have involved U.S. forces in various missions. These missions include SSC, SASO, MOOTW, and civil disturbances where the ROE are inherently restrictive. NLW are designed to address the special requirements of low-intensity environments. They provide tools that enable the individual, team, or unit, to anticipate and/or respond to provocation with more appropriate means than through the use of deadly force. Essentially, NLW provide U.S. forces with a near-term capability for effectively controlling the nontraditional battlefield, within the constraints levied by ROE, by reducing the risk to friendly forces, mitigating casualties, and minimizing collateral damage.

2. System Description

a. NLW Capability Set. The services have, and are continuing to procure, versatile packages of NLW comprised of commercial off-the-shelf (COTS) and government off-the-shelf (GOTS) equipment and munitions. The government is also exploring advanced technologies for future procurements. In addition to addressing contingency requirements, the services' NLW capability sets provide limited sustainment training ammunition and equipment.

b. NLW Capability Set—Component Categories. Each service determines the contents of its own capability set. Examples of the services current configurations are shown in Appendix E. The NLW capability set components can be divided into four distinct categories: personnel protectors, personnel effectors, mission enhancers, and training devices. The following examples are not all-inclusive.

(1) Personnel Protectors. Personnel protectors include items such as face shields, riot shields, and shin guards that protect the individual from blunt trauma injuries inflicted by thrown objects, clubs, etc.

(2) Personnel Effectors. Personnel effectors include items such as riot batons, malodorants, stingball grenades, pepper sprays, and other kinetic rounds (e.g. sponge grenades) designed to discourage, disorient, or incapacitate individuals or groups.

(3) Mission Enhancers. Mission enhancers include items such as bullhorns, spotlights, caltrops, isotropic radiators (stun/flash), and barriers/entanglers. These items are designed to facilitate target identification and crowd control, and to limit personnel and vehicular movement.

(4) Training Devices. Training devices include items such as training suits, training rounds, training batons, and inert pepper sprays. They are designed to facilitate realistic hands-on scenario training in preparation for operations.

c. System Relationships and Inter/Intraoperability. The NLW capability set is designed to provide an operating force (ranging from squad through battalion-sized units [see Appendix E]) with tactically deployable nonlethal equipment. Fielding of this capability set enhances the operation of systems and equipment already fielded and planned within the joint community. The NLW capability set components are compatible with military equipment that is already fielded or planned for future fielding. The components of the NLW capability set are supported within current military logistic capabilities.

d. Augmentation of Organic Equipment. The operational intent of the capability set is to augment an operating force with the required nonlethal equipment to enhance its force projection capability. The NLW capability set is additional equipment that must be incorporated into lift and load planning.

3. Organization

Organization/organizational structure is not anticipated to change as a result of NLW capability sets being added to the inventory.

a. System Location/Distribution. The United States Army (USA), United States Marine Corps (USMC), and United States Navy (USN) have procured and fielded capability sets to the operating forces, security forces, bases, and stations. All services have current plans to field additional capability sets as required. At the time of this printing, the United States Air Force (USAF) has identified its capability sets to field in the near future. The United States Coast Guard (USCG) also intends to field NLW.

b. NLW Capability Set Composition. Appendix E describes the current and proposed capability sets.

c. Personnel Requirements. To date, no change or increase in military specialty codes have been determined in order to field, operate, and maintain the NLW capability set. Each service determines its own training personnel requirements.

d. Mission Requirements/Tactical Structure.

(1) The mission requirements and tactical structure pertinent to the employment of the NLW will vary depending on the type of contingency and subsequent task organization of the employing unit. Mission requirements, derived from commander's guidance, should be satisfied through the appropriate combination of lethal and nonlethal capabilities.

(2) NLW provide commanders with a means of shaping their area of operations (AO) short of employing deadly force. However, it is imperative that U.S. forces retain the means and ability to employ lethal force in the event that nonlethal force fails to achieve desired outcomes. For example, NLW employment may provoke a lethal response from the adversary. Without the imminent threat of lethal force, nonlethal munitions will prove to be less effective and possibly even counterproductive. More importantly, denying U.S. forces a lethal capability would place them at unacceptable risk. If nonlethal force is employed without the threat of lethal force, the belligerent may opt to exploit a perceived vulnerability. In this case, the use of NLW becomes counterproductive, because it serves only to aggravate the threat and causes the situation to deteriorate.

Chapter III

NLW CAPABILITY REQUIREMENTS

1. Applicability Across the Range of Military Operations

a. U.S. forces conduct military operations across a large spectrum. This spectrum spans from humanitarian assistance and peacekeeping to strategic nuclear exchange. Since the end of the Cold War, the U.S. military has been increasingly involved with operations at the lower end of the spectrum. These operations focus on deterring war, resolving conflict, supporting civil authorities and promoting peace. NLW may be useful across many categories of these operations. The range of military operations includes various levels of combat "intensity," and NLW provide one more force option to deal with the varying intensity. MOOTW may be found across the range of military operations and often include situations where noncombatants are mingled with the adversaries or involved in acts of violence, such as rioting or looting. In such circumstances, NLW provide commanders the flexibility to influence the situation favorably with reduced risk of noncombatant fatalities and collateral damage. NLW also provide a means by which to ascertain the intent of individuals/belligerents. Some characteristics of operations that present the opportunity to use NLW are—

(1) Close contact between friendly forces and noncombatant civilians.

(2) The presence of paramilitary or irregular forces.

(3) Armed civilian gangs, unlawful combatants, or militia.

b. These characteristics present a real but ill-defined threat. Under such circumstances, the commander's judgment or ROE may restrict using deadly force for other than self-defense. NLW may provide U.S. forces flexibility to resolve situations where the use of deadly force is restricted. These weapons provide flexibility by—

(1) Allowing U.S. forces to apply measured military force.

(2) Reducing risk of serious noncombatant causalities.

(3) Accomplishing the mission.

(4) Providing force protection.

(5) Retaining the initiative.

(6) Allowing U.S. forces to remain relevant and capable in increasingly complex and chaotic environments.

c. The need to reduce the risks of serious injury to personnel is not limited to crowd control scenarios. In MOUT, some of the local civilian populace may remain in an urban area in the midst of battle. The traditional solution to such challenges has been restricting ROE. Nonlethal capabilities may provide commanders the flexibility to adapt a fluid approach to MOUT, which allows their subordinates the freedom of action to employ appropriate levels of measured military force to accomplish their mission.

d. Another example of NLW applications in MOOTW is in peace enforcement. In these operations, NLW are used in an area denial role to separate belligerents without resorting to deadly force. They are also used to support humanitarian relief organizations in protecting food and medical distribution sites. The presence of NLW will not limit a commander's inherent authority and obligation to use all necessary means available and to take appropriate action in self-defense.

2. Core Capabilities

Core capabilities are those fundamental competencies that enable U.S. forces to achieve desired operational objectives. NLW provide a flexible means to protect friendly forces, high-value assets, and influence the actions of potential adversaries and/or noncombatants. These goals can be achieved without resorting to lethal force and in a manner that may minimize collateral damage. The core capabilities associated with nonlethal effects fall into three major categories as stated within the Joint Mission Area Analysis (JMAA) for Nonlethal Weapons (approved December 2000 by the Joint Requirements Oversight Council): counterpersonnel, countermateriel, and countercapability.

a. Counterpersonnel Capabilities. The DOD Joint Concept for Nonlethal Weapons concept paper, 5 January 1998, describes required nonlethal counterpersonnel capabilities for U.S. forces as follows:

(1) "The application of military force to accomplish a mission with reduced risk of fatalities or serious casualties among noncombatants—or even, in some instances, among enemy forces." There are several specific nonlethal counterpersonnel capabilities to be considered. These include, but are not limited to, the means to influence the behavior and activities of a potentially hostile crowd and the capability to bring a mob engaged in a riot under control. While there are many similarities in these two capabilities, each involves unique challenges, which may require radically different solutions.

(2) "Capability to incapacitate personnel." For the purposes of this document, "incapacitation" is achieved when the weapon's effects render personnel unable to physically or mentally (real or perceived) act in a hostile or threatening manner. In keeping with the guiding principles for NLW, this incapacitation should be reversible through the passage of time. The focus will be on capabilities that affect both individuals and groups of people.

(3) "Capability to deny personnel access to an area (land, sea, or air)." NLW can provide new possibilities for barrier planning in any type of military operation, including systems that produce physical or mental discomfort to those who enter the denied area.

(4) "Capability to clear facilities and structures of personnel." The Joint Concept for NLW states this capability "will facilitate MOUT by reducing the risks of noncombatant casualties and collateral damage while simultaneously minimizing the advantages accruing to an enemy defending a built up area." This capability is applicable to all clearing missions that may be conducted (e.g., Maritime Interception Operations [MIO]) and could assist in providing enhanced force protection.

(5) "Capability to seize personnel, which could include some combination of technologies inherent in other core capabilities, for example, incapacitation and entangling devices (which may aid in seizing personnel)." This capability is intended to augment lethal means used to capture specified individuals, such as enemy combatants or persons who are inciting a mob or crowd to violence.

b. Countermateriel Capabilities. The Joint Concept for Nonlethal Weapons describes countermateriel capabilities as follows:

(1) Rendering equipment and facilities unusable without complete destruction by attacking only weapons of war and supporting infrastructure while minimizing risk to noncombatants. This could enhance operations by defusing potentially volatile situations where conventional military means might otherwise prove counterproductive.

(2) Disabling or denying the use of vehicles, vessels, and aircraft entry into targeted areas or access within an AO. This capability may include physical barriers that impede or stop movement. Alternative systems may alter combustion properties of fuels

and viscosity of lubricants, embrittle or decay materials, attack tires, gaskets, or hoses, or act as adhesives to fuse metal parts. The requirement also includes the capability to facilitate boarding and clearing operations.

c. Counter Capability Capabilities. For U.S. forces counter capability capabilities are described within the JMAA for NLW as follows:

(1) "Disable or neutralize facilities and systems." This includes a nonlethal capability for disabling/neutralizing electrical generating facilities, command, control, communications, computers, intelligence, surveillance, and reconnaissance (C4ISR) systems, Integrated Air Defense Systems (IADS), weapons systems, optical sensors, electrical sensors, and navigation capabilities with such controlled effects as to allow for selective, precise engagement. Other needs include deceiving reconnaissance with illusions of barriers, obstacles, and/or forces, and concealing/covering friendly activities or movements, particularly within complex urban terrain.

(2) "Deny the use of weapons of mass destruction (WMD)." Mission needs include rendering a WMD inoperative; containing the potential release of deadly agents/contaminants; and preventing or neutralizing the production, storage, deployment (transport), employment, and delivery of WMD. Nonlethal capabilities in combating WMD should be designed to permit engagement within a populated or sensitive terrain, prior to the initiation of armed conflict.

3. Limitations

The limitations of NLW are effectiveness, perception, and legality.

a. The effectiveness of NLW is dependent on factors such as motivation, age, environmental and human factors, range of engagement, and numerous others. NLW can be lethal if improperly applied. Furthermore, countermeasures for thwarting virtually all nonlethal options are usually apparent, quickly learned, and readily available. Because they are not intended to kill, nonlethal options teach adversaries what to avoid in the future. Small unit commanders must prepare to keep one step ahead of belligerents.

b. The perceptual limitation is that personnel may misunderstand the appropriate applications of NLW across the range of military operations. The incorrect perception that NLW will allow wars and MOOTW to be prosecuted without casualties may lead to conflicting expectations between political and military leaders. These conflicts can result in U.S. forces being vulnerable to misplaced or inconsistent ROE. All leaders, political and military, involved in planning and executing military missions must understand there are no "nonlethal operations."

c. Some proposed NLW might be forbidden by law or policy. Accordingly, it is essential that all NLW developments be evaluated by appropriate authorities to ensure they comply with the law of war, U.S. law, and U.S. treaty obligations. Use of NLW containing chemical agents, for instance, must be evaluated in the context of the Chemical Weapons Convention. Also, using RCA in an armed conflict requires Presidential approval.

Chapter IV

TRAINING

1. Fundamental Concepts and Training Requirements

The successful accomplishment of any operation in which nonlethal measures are employed requires extensive preparation, of which individual, unit, medical support, and unit training are vital parts. Training should be designed to give individuals an understanding of the entire subject area and enable them to function efficiently as members of a unit. This training must be intensive and realistic. Training with the NLW capability set, and other NLW, is critical to employ these weapons with confidence. The training flow should be—train leaders, train instructors, train unit, exercise. Qualified instructors should conduct initial unit training. Unit sustainment training should be conducted per individual service and unit requirements.

2. Training of Instructors

a. Verbal communication skills, open-hand control techniques, RCA, effects of impact weapons, and civil disturbance tactics, techniques, and procedures (TTP) are areas trained to standard at the Interservice Nonlethal Individual Weapons Instructor Course (INIWIC) held at the Marine Corps Detachment, Fort Leonard Wood, Missouri. INIWIC is an Interservice Training Review Organization (ITRO) course, and quotas are available to any service through normal training reservation channels. A course outline is provided in Appendix C. This is the only formal DOD NLW instructor-training course. The training requirements outlined in the INIWIC represent the minimum training standards to include in developing NLW and tactics instructors. Additionally, the United States Army Military Police School (USAMPS) has developed a training support package that includes Army-specific TTP for small units (company level and below) in using NLW capabilities. The reference for this is the U.S. Army's website at: www.wood.army.mil.

b. The INIWIC is a "train-the-trainer" program to provide commanders NLW instructors within their command. Furthermore, the military services are not precluded from developing their own training based upon the INIWIC curriculum. The commander can then tailor the NLW instruction to fit mission needs. Establishing the service-level INIWIC serves as an assurance measure, giving the commander a quantifiable means of certifying his nonlethal instructors.

3. Nonlethal Training Topics for Deploying Units

Each service and unit will have unique training requirements based on mission type, unit size, deployment area, etc. As a minimum, the training instructor should cover the following topics in any training for units deploying with nonlethal capabilities. They are part of the INIWIC curriculum.

NOTE: U.S. Army training standards are contained in Army Regulation (AR) 350-38, Training Devices Policies and Management, 15 October 1993; USMC training standards for NLW are contained in MCO 1510.112, Individual Training Standards for Nonlethal Weapons, 27 May 1998.

---

a. Force Continuum. This is an introduction to the federal force continuum model as outlined by MCO 5500.6F, Arming of Law Enforcement and Security Personnel and the Use of Deadly Force; AR 190-14; AFI 31-207, Arming and Use of Force by Air Force Personnel (1 September 1999) (or other service orders). Identifying the proper levels of force, understanding levels of resistance, and how nonlethal technologies affect the force continuum are the minimum topics the instructor should cover.

b. Crowd Dynamics/Crowd Control. This is an introduction to the differences between crowds, mobs, and riots. This course teaches the student basic crowd control techniques that will be easily applied to various situations. The student should be familiarized with tactics and techniques found in field manual (FM) 19-15, MCWP 3-11.1, Rifle Company, TC 90-1, and Air Force manual (AFMAN) 31-201V6, Civil Disturbance (1 April 2002), but will also consider nontraditional and small unit application.

c. Communication Skills. This is an introduction to the stages of conflict management, verbal aggression, nonverbal communication, physical aggression, physiological diversions, and proper mental conditioning. The first NLW skill an individual must have is the ability to communicate. General A. M. Gray, former Commandant of the Marine Corps, once said, "A warrior's most formidable weapon is his mind." The results of an individual's mental processes are evident through actions and words, and the words and communication skills used in dealing with people are truly the first line of nonlethal options.

d. Oleoresin Capsicum (OC) Aerosol and other RCA Training. This is an introduction to the uses of OC/pepper spray and other RCA. The student should have an appreciation for decontamination requirements, storage, and shelf life of OC/RCA. The instructor will also teach legal and tactical considerations. INIWIC certified instructors should teach existing training programs that include live agent training.

e. Open-Hand Control. This is an introduction to pressure point control techniques, unarmed self-defense measures, weapon retention techniques, and other submission/restraint/search techniques. The instructor should cover handcuffing/flex-cuffing for contact team members.

NOTE: USMC plans to replace the current open-hand defensive tactics training with the USMC Martial Arts Program.

---

f. Impact Weapons. This is an introduction to using the rigid straight baton, collapsible straight batons, side handle batons, and/or riot control batons.

g. ROE/Law of Armed Conflict (LOAC). This is an introduction to the ROE/LOAC and their relationship. (Refer to DODD 5100.77, AR 27-1, and NWP 1-14M).

h. Nonlethal Munitions—Employment and Tactics. This subcourse is an introduction to the capabilities and employment of available NLW. NLW provide additional unit force protection capabilities. Introduction to tactics is outlined in Chapter 5. Appendix D is a sample unit training schedule covering the topics listed above. U.S. Army Training and Doctrine Command (TRADOC) has developed training support packages that will include company level and below TTP.

4. Leader Training

a. An understanding of the topics covered in the INIWIC curriculum is essential. However, commanders/leaders should also possess an in-depth understanding of the ROE, their implementation, as well as the politics, culture, and demographics of the area to which they are deploying.

b. Commanders should involve their medical personnel, public affairs (PA) representatives, civil affairs (CA), psychological operations (PSYOP), staff judge advocate

(SJA), engineers, explosive ordnance disposal (EOD), and military police (MP) in training exercises when possible. Involvement by these special staff sections will greatly enhance the quality and realism of the training. Medical personnel will better understand OC decontamination and the possible injuries they may encounter as a result of NLW employment. Because of the international attention that NLW receive, all personnel will benefit from media training by PA personnel. CA and PSYOP personnel can provide area-specific cultural, ethnic, and political briefs to enhance awareness of the challenges within the AO. SJA personnel can clarify complex ROE questions and assist in training on ROE-related issues. Additionally, supply and logistic staff members also require training in NLW maintenance and environmental issues unique to the equipment and munitions.

5. Unit Training

The INIWIC is primarily designed to produce instructors who will train individuals in using and employing NLW. Collective unit tactical training is the unit's responsibility; however, a NLW sample schedule based on INIWIC training is provided in Appendix D. Units should conduct collective training before deploying with NLW. Unit standard operating procedures (SOP) and tactics may require adjustment to enable the employment of NLW and new equipment for the success of specific missions. All collective training should include nonlethal options and decision making. Whenever NLW are employed during training, the instructor must also conduct lethal capability training. Troops must understand the integration of nonlethal and lethal capabilities and always be prepared to protect themselves, as required, with lethal force. It is impossible to accurately predict the response of an individual or crowd.

6. Training Disclaimer

Initial unit training should not be conducted without qualified instructors. Non-type classified and COTS nonlethal munitions cannot be fired without Army Materiel Command (AMC), Army Training Support Center (ATSC), Crane Naval Weapons Center, or Air Force Munitions Center type classifying the ammunition. The instructor should conduct unit sustainment training in accordance with (IAW) service guidelines.

7. Additional Training

a. Additional training is left to the discretion of the unit but should be encouraged in order to add credibility to the training package. Some recommended training certifications for instructors include, but are not limited to –

(1) Management of Aggressive Behavior (MOAB)
R.E.B. Training International, Inc.
PO Box 845, Stoddard, NH 03464

(2) Oleoresin Capsicum Aerosol Training (OCAT)
R.E.B. Training International, Inc.
PO Box 845, Stoddard, NH 03464

(3) Monadnock Defensive Tactics System (MDTS)
Training calendar available at the Monadnock website:
http://www.batons.com/traincal.html

b. These courses add viable credentials to the instructors as perceived by civilian and military alike. In addition, the instructor can issue certification cards for the area which an individual has been certified. Certifications in various areas offers additional protection in the event of a lawsuit or claim against the government by ensuring the individuals that have been trained within your unit are completely capable.

8. Summary

NLW and tactics training are not meant to be stand-alone training—units must anticipate the requirement for NLW in their missions and train accordingly. NLW provide commanders with a graduated response in the force continuum.

Chapter V

EMPLOYMENT

1. Background

a. A military unit deployed to an area will normally conduct operations IAW current doctrine. The environment may contain both noncombatants and combatants. Nonlethal capabilities provide the military force with tools to engage adversaries and minimize collateral damage. Fundamental to employing NLW is a thorough understanding of the force continuum. As stated before, nonlethal capabilities provide a wider range of options that augment traditional means of deadly force but do not replace them. Nonlethal capabilities allow a commander the flexibility to increase and decrease the amount of force applied to accomplish a mission. The force continuum is generally continuous and seamless, yet a careful examination reveals five broad categories:

(1) Threats. The force continuum begins with threats. These can be either implied or expressed. An implied threat may be manifested through presence by forming into riot-control formations, fixing bayonets, or donning protective masks. An expressed threat occurs when a commander makes known the consequences of defiance.

(2) Denial options. The equipment used in denial options are usually placed in highly visible areas and often marked with placards. Examples include concertina, caltrops, barbed wire, or other obstacles.

(3) Munitions that cause physical discomfort/incapacitate. Munitions that cause physical discomfort, but fall short of inflicting trauma, include flash bangs, RCA—tear gas (O-Chlorobenzyl-Malononitrile [CS]), and OC. Although the discomfort or injury may be substantially less than impact munitions, employing these munitions requires an appropriate use of force assessment. Factors such as training, discipline, prejudices, emotions, and judgment all play a part in their application.

(4) Munitions that inflict blunt trauma. Munitions that inflict blunt trauma will inflict painful or debilitating injuries. These munitions constitute the upper end of nonlethal options within the force continuum. Examples include foam batons, stingball grenades, beanbags, sponge projectiles, and rubber ball rounds.

(5) Lethal munitions. Lethal weapons are at the highest level of force continuum. Lethal options should always be regarded as part of the force continuum and not as a separate option altogether. ROE, along with judgment and experience, helps on-scene commanders decide when using lethal force is authorized and appropriate.

b. Nonlethal capabilities include using existing systems that were not designed specifically as NLW. Nonlethal capabilities can minimize fatalities, permanent injury to personnel, and undesired damage to property, facilities, and the environment. Examples include PSYOP, CA, military working dogs, and PA. However, this chapter limits discussion primarily to using weapons and components explicitly designed as part of a NLW capability set. This focus is not intended to limit the commander's option to use whatever means are available to accomplish the mission without unnecessary lethality.

2. Mission Planning Factors

Incorporating NLW into a plan does not change the fundamental principles of mission planning. Commanders should use NLW as an operational enhancement. The first concern of a commander is successfully completing the mission. ROE are a critical consideration in the commander's planning and guide him in determining the appropriate use of nonlethal and lethal force. The availability of NLW does not relieve a force from its obligation to defend itself. Therefore, it must be equipped and trained with both lethal and nonlethal means.

3. NLW Tasks

a. NLW tasks were derived from the combatant commanders and service mission needs statements and are codified in the JMAA for Nonlethal Weapons. These tasks are grouped under three core capabilities: counterpersonnel, countermateriel, and countercapability.

(1) Counterpersonnel:

(a) Control crowds.

(b) Incapacitate personnel.

(c) Deny an area to personnel.

(d) Clear facilities/structures of personnel.

(2) Countermateriel:

(a) Deny an area to vehicles, vessels, and aircraft.

(b) Disable/neutralize vehicles, vessels, aircraft, and equipment.

(3) Countercapability:

(a) Disable or neutralize facilities and systems.

(b) Deny use of WMD.

b. Although not all encompassing, the following list of employment considerations is designed to assist commanders and staffs in formulating plans for use of NLW:

(1) General considerations:

(a) Understand and define ROE.

(b) Commander develops identification of friend or foe (IFF) procedures.

(c) Distribute NLW munitions to troops in advance.

(d) NLW options require lethal overwatch fire.

(e) Ensure area is well lighted to include perimeter.

(f) Employ night vision devices outside the perimeter to identify personnel.

(g) Utilize appropriate media to inform civilians to stay away from denied areas.

(h) Enforce positive access control into perimeter (personnel/vehicles).

(i) Ensure posted signs are easily understood (symbols/pictures). If interpreter support available, post warning signs in native language.

(j) Use entanglements such as caltrops, tanglefoot, barbed wire, and concertina/razor tape as these disrupt/slow intrusion attempts.

(k) Employ sensors, if available.

(l) Employ aerial or naval pickets.

(2) Counterpersonnel considerations:

(a) Employing kinetic rounds at less than 15 feet/5 meters may result in fatal outcomes.

(b) Target area at 15 to 40 feet should be center mass for blunt munitions.

(c) Head shots may be lethal.

(d) Kinetic rounds will not be skip fired.

(e) Kinetic rounds are effective against selected individual targets and can be used to disperse individuals.

(f) Considering differing rates of fires; the M203 and shotgun are complementary in a nonlethal role, depending upon the munitions used and their effective ranges.

(g) Employ riot batons only after proper training.

(h) Use RCA to disrupt/disperse crowds.

(i) Use RCA markers (RCA with a nontoxic, water-soluble marking dye) to mark agitators for future identification and apprehension.

(j) RCA greatly reduce visibility of control force.

(k) Closely control the employment of RCA.

(l) Disperse RCA to face and eyes for full effect.

(m) Employ optics with designated marksmen to identify mob leaders.

(n) Searchlights dazzle/disorient individuals (night only).

(o) Use bullhorn for communicating with crowd and to control formations.

(p) Employ recovery/apprehension/snatch teams to apprehend agitators.

(q) Use flex cuffs for agitators.

(r) Prepare disposition plan for cleared personnel.

(s) Isolate area to prevent reinforcements.

(t) Consider future friendly use of cleared facility.

(u) Prepare security plan for cleared facility.

(v) Channel passive traffic away from formations.

(w) Maintain rapid reaction force/quick reaction force (RRF/QRF).

(3) Countermateriel/countercapability considerations:

(a) Use marking agents to mark a vessel for later interception.

(b) Consider materiel handling equipment required to evacuate vessel/airframe.

c. Selected NLW tasks are further addressed in Tables V-1 through V-7. The tables are not intended to be all-inclusive but merely a stimulus to planning. The equipment listed is taken from the currently available systems designed for nonlethal purposes. The user should not limit his nonlethal options to this specific equipment and should be aware that there are many existing and emerging systems and potential uses of current technology and common objects to achieve the desired mission end state.

Table V-1. Crowd Control

Tasks	*NLW Planning Factors*	*Equipment*	*Employment Considerations*	*Notes*
Crowd control	Administration/ Personnel • SJA / claims planning • Chaplain affairs • Casualty notification • Detainee disposition plan • PAO/media planning • Information dissemination	Support Equipment • Riot face shield • Full length riot shield • Expandable riot baton • Optics • Portable bullhorn • Vest	• Distribute to troops in advance • Employ riot baton only after proper training • Employ optics with designated marksman (DM) to identify (ID) mob leaders • Use bullhorn for communicating with crowd and to control formations	• Strictly define ROE • NLW options require lethal cover fire
Crowd control	Intelligence • Intelligence preparation of the battlespace (IPB) • Background on country and culture • Route/map recon • Sensor emplacement for advance notification • Information on Mob leaders • Patrols and listening post (LP)/observation post (OP) • Determine reason for the crowd	Kinetic Rounds • 12-gauge (ga) bean bag round • 12-gauge rubber bullet • 12-gauge launching cartridge • 40-mm rubber baton • 40-mm wooden baton **NOTE:** USMC no longer uses wooden baton. • 40-mm stinger grenade • Stun grenade • Flash bang	• Do not employ kinetic rounds at less than 15 feet due to possible fatal outcome • Target area at 15 to 40 feet should be center mass • Head shots are not acceptable • Kinetic rounds will not be skip fired • Kinetic rounds effective against selected targets and to disperse individuals • M203 and shotgun are nonlethal complementary; M203 has low rate of fire; area target. Shotgun has high rate of fire; point target	

Table V-1. Crowd Control

Tasks	*NLW Planning Factors*	*Equipment*	*Employment Considerations*	*Notes*
Crowd control	Operations • Tactical Integration of NLW and lethal fires • Fire Support Element (FSE) plans to include RCA • Allocation of NLW munitions • Integration of NLW fires and maneuver • Engineer barrier planning • Understand mission (deny access/disperse/ monitor)	RCA • OC Dispenser • Team OC dispenser • High volume dispenser	• Disrupt/disperse crowds • Disperse RCA to face and eyes for full effect • Employ snatch teams to apprehend agitators	
Crowd control	Logistics • Distribution of NLW supplies • Security of rear area facilities • Resupply of NLW munitions • Procurement of medical supplies • Hospital planning • Medic planning for tactical units	Riot Control • Xenon Searchlight • Flex cuffs • RCA markers	• Searchlights dazzle/disorient individuals (night only) • Flex cuffs used for agitators • RCA markers are used to mark agitators for future ID	

Table V-2. Incapacitate Personnel

Task	*NLW Planning Factors*	*Equipment*	*Employment*	*Notes*
Incapacitate personnel	Administration/Personnel • SJA/claims planning • Media plan	Support Equipment • Riot face shield • Full length riot shield • Expandable riot baton • Rifleman's combat optic • Vest	• Distribute to troops in advance • Do not employ riot baton to the head • Employ optic with DM to ID mob leaders and cover with lethal fire	• Commander needs to define incapacitation mission • Take a balanced approach to public dissemination of NLW so counter measures cannot be rapidly developed • Cover NLW options by lethal force
Incapacitate personnel	Intelligence • ID types of insurgents and demonstrators • ID causes and factions • ID mob ringleaders	Kinetic Rounds • 12-gauge bean bag round • 12-gauge rubber bullet • 12-gauge launching cartridge • 40-mm rubber baton • 40-mm wooden baton **NOTE:** USMC no longer uses wooden baton. • 40-mm stinger grenade • Stun grenade • Flash bang • M203 and shotguns are nonlethal complementary. M203 has a low rate of fire, area target. Shotgun has a high rate of fire, point target	• Do not employ kinetic rounds at less than 15 feet due to possible fatal outcome • Target area at 15-40 feet should be center mass • Head shots are not acceptable • Kinetic rounds will not be skip fired • Kinetic rounds effective against selected targets and to disperse individuals • M203 and shotguns are nonlethal complementary. M203 has a low rate of fire, area target. Shotgun has a high rate of fire, point target	

Table V-2. Incapacitate Personnel

Task	*NLW Planning Factors*	*Equipment*	*Employment*	*Notes*
Incapacitate personnel	Operations • Prepare and confirm ROE for NLW • Integration of NLW and deadly force • Allocate NLW and designate unit use • Integrate NLW fires and maneuver	RCA • OC dispenser • Team OC dispenser • High-volume • OC dispenser	• Disrupt/disperse crowds • Disperse RCA to face and eyes for full effect	
Incapacitate personnel	Logistics • Distribute NLW munitions • ID special storage requirements • Resupply NLW munitions	Riot Control • Xenon • searchlight • Flex cuffs • Chemical markers	• Searchlights dazzle/disorient individuals (night only) • Use flex cuffs for agitators • Use RCA markers to mark agitators for incapacitation	
Incapacitate personnel	Civil Affairs • Collect civilian intelligence • Care for injured personnel • Detain civilian personnel			

Table V-3. Deny Area to Personnel

Task	*NLW Planning Factors*	*Equipment*	*Employment*	*Notes*
Deny area to personnel	Administration/ Personnel • SJA/claims planning • Media plan	Support Equipment • Area lights • Area sensors • Night vision • Access control • Warning signs written in host nation language	• Ensure area is well lighted to include perimeter • Employ night vision devices outside the perimeter • Positively ID accessed personnel • Employ warning signs in native language	• Utilize appropriate media to warn civilians away from denied area

Table V-3. Deny Area to Personnel

Task	*NLW Planning Factors*	*Equipment*	*Employment*	*Notes*
Deny area to personnel	Intelligence • Terrain and map study • Threat ID • Sensor emplacement around area • IPB			
Deny area to personnel	Operations • Prepare and confirm ROE • Integrate NLW and deadly force • Allocate NLW and designate unit use • Determine scope of mission • Determine access procedures • Use of area (friendly or total denial)			
Deny area to personnel	Logistics • Distribute NLW munitions • ID special storage requirements • Resupply NLW munitions • Security of rear area facilities.			
Deny area to personnel	Civil Affairs • Collect civilian intelligence	Entanglements • Caltrops • Tanglefoot • Barbed wire • Foams • Concertina/ razor tape • Stinger spike system	• Disrupt/slow intrusion attempts • Maintain suitable reaction force • Channel passive traffic away • Use foam to cover caltrops	

Table V-4. Deny Area to Vehicles

Task	*NLW Planning Factors*	*Equipment*	*Employment*	*Notes*
Deny area to vehicles	Administration/ Personnel • SJA/claims planning • Media plan	Support Equipment • Area lights • Area sensors • Night vision • Access control • Warning signs written in host nation language	• Ensure area is well lighted to include perimeter • Employ night vision devices outside the perimeter • Positively ID accessed vehicles	• Commander needs to define denial mission • Utilize appropriate media to warn civilians away from denied area
Deny area to vehicles	Intelligence • Terrain and map study • Threat ID • Sensor emplacement around area • IPB • Traffic analysis			
Deny area to vehicles	Operations • Prepare and confirm ROE • Allocate NLW and designate unit use • Determine scope of mission • Determine access procedures • Determine MP/ security forces requirements • Use of area (friendly or total denial)			• Commanders must consider flying operations when employing RCA. Area(s) should be reusable in short order.
Deny area to vehicles	Logistics • Distribute NLW supplies • Be prepared to evacuate stalled vehicles			
Deny area to vehicles	Civil Affairs • Collection of civilian intelligence	Entanglements • Caltrops • Tanglefoot • Barriers • Organic equipment • Barbed wire • Foams	• Disrupt/slow intrusion attempts • Maintain suitable reaction force • Channel passive traffic away • Use foam to cover caltrops	• Commanders must consider flying operations when employing NLW. NLW should not hinder flying operations.

Table V-5. Deny/Neutralize Vehicles, Aircraft, Vessels, and Facilities

Task	*NLW Planning Factors*	*Equipment*	*Employment*	*Notes*
Deny/ neutralize vehicles, aircraft, vessels, facilities	Administration/Personnel • SJA/claims planning • Media plan	Support Equipment • Sensors to track approaching vehicles, aircraft, and vessels • Night vision equipment • Materiel handling equipment • Portable bullhorn	• Employ ground and water sensors, if required • Employ aerial or naval pickets • Use materiel handling equipment to evacuate vessel	• Commander needs to define neutralization mission • Element of surprise remains a critical factor in mission accomplishment • Commander needs to develop IFF procedures
Deny/ neutralize vehicles, aircraft, vessels, facilities	Intelligence • Sensor emplacement if required • Threat ID • IPB • Analyze target area			
Deny/neutralize vehicles, aircraft, vessels, facilities	Operations • Prepare and confirm ROE • Integrate NLW and deadly force • Allocate NLW and designate unit use • Determine scope of mission • Determine access procedures • Determine boarding procedures (vessels and aircraft) • Determine apprehension procedures • Determine decontamination (DECON) procedures	RCA • Marking agent	• Use marking agents to mark a vessel for later interception	

Table V-5. Deny/Neutralize Vehicles, Aircraft, Vessels, and Facilities

Task	*NLW Planning Factors*	*Equipment*	*Employment*	*Notes*
Deny / neutralize vehicles, aircraft, vessels, facilities	Logistics • Distribute NLW munitions • ID special storage requirements • Resupply NLW munitions • Determine vessel/aircraft disposition	Riot Control • Xenon searchlight	• Use RCA to incapacitate crew members • Searchlights can disorient personnel or illuminate target	• Commanders must consider flying operations when employing RCA • Area(s) should be reusable in short order.
Deny / neutralize vehicles, aircraft, vessels, facilities	Civil Affairs • Collect civilian intelligence • Handle displaced civilian personnel	Entanglements • Caltrops • Foams • Speed bumps • Nets • PVAB	• Use caltrops to disable vehicle • Use combination of foam and caltrops to preserve the element of surprise • Channel passive traffic away • Use nets to foul props	
Deny/neutralize vehicles, aircraft, vessels, facilities	Medical • Treating injured civilian personnel			

Table V-6. Seize Personnel

Task	*NLW Planning Factors*	*Equipment*	*Employment*	*Notes*
Seize personnel	Administration/ Personnel • SJA/claims planning • PAO media plan	Support Equipment • Optics	• Employ optics with DM to ID personnel to be seized • Cover personnel employing nonlethal force	• Commander needs to define seizure mission and strictly enforce ROE • Element of surprise is critical for mission accomplishment • Speed is critical • Seizing personnel can be surgical with special operations personnel or conducted within a crowd to seize an agitator
Seize personnel	Operations • Prepare and confirm ROE • Integrate NLW and deadly force • Allocate NLW and designate unit use • Integrate NLW fires and maneuver • Breaching and assault planning • Determine holding area/processing center	RCA • OC dispenser • CS	• Use RCA to incapacitate personnel • Spray into eyes and face for full effect	
Seize personnel	Logistics • Distribute NLW munitions • ID special storage requirements • Resupply NLW munitions • Plan transportation	Riot Control • Xenon searchlight • Flex cuffs • RCA markers	• Searchlights dazzle/disorient individuals • Use flex cuffs for apprehension • Use RCA markers to mark agitators for seizure	

Table V-6. Seize Personnel

Task	*NLW Planning Factors*	*Equipment*	*Employment*	*Notes*
Seize personnel	Civil Affairs • Collect civilian intelligence • Care for injured personnel • Detain civilian personnel			
Seize personnel	Medical • Medical evacuation (MEDEVAC) planning			

Table V-7. Clear Facilities of Personnel

Tasks	*NLW Planning Factors*	*Equipment*	*Employment Considerations*	*Notes*
Clear facilities of personnel	Administration/ Personnel • SJA/claims planning • Chaplain affairs • Casualty notification • PAO media plan	Support Equipment • Riot face shield • Full-length riot shield • Expandable riot baton • Optics • Portable bullhorn	• Distribute to troops in advance • Do not employ riot baton to the head • Employ optic with DM to ID occupant leaders and provide lethal cover fire • Use bullhorn used for communicating with occupants	• Surprise is a critical element for mission accomplishment • Commander must strictly define ROE • Due to close quarter tactics, the force must be instantly ready to utilize deadly force if required

Table V-7. Clear Facilities of Personnel

Tasks	*NLW Planning Factors*	*Equipment*	*Employment Considerations*	*Notes*
Clear facilities of personnel	Intelligence • IPB • Background on facility, access, power plant • Route/map recon • Background info on building occupants • Determine any outside support for building occupants	Kinetic Rounds • 12-gauge bean bag round • 12-gauge rubber bullet • 12-gauge launching cartridge • 40-mm rubber baton • 40-mm wooden baton **NOTE:** USMC no longer uses wooden baton. • 40-mm stinger Stun grenade • Flash bang	• Do not employ kinetic rounds at less than 15 feet due to possible fatal outcome • Targeted personnel at 15 to 40 feet should be center mass • Head shots are not acceptable • Kinetic rounds will not be skip fired • Kinetic rounds effective against selected targets and to disperse individuals • Stun grenade effects can be amplified in buildings • Visibility will be reduced once kinetic rounds are employed • M203 and shotguns are nonlethal complementary. M203 has a low rate of fire, area target. Shotgun has a high rate of fire, point target	• Using RCA greatly reduces visibility • Use of RCA is restricted to certain circumstances; verify that proper authority has been given before use

Table V-7. Clear Facilities of Personnel

Tasks	***NLW Planning Factors***	***Equipment***	***Employment Considerations***	***Notes***
Clear facilities of personnel	Operations • Tactically integrate NLW and lethal fires • FSF plans to include RCA • Allocate NLW munitions • Integrate NLW fires and maneuver • Determine assault and breaching plan • Plan for isolation of facility	RCA • OC dispenser • Team OC dispenser • High-volume dispenser	• Disrupt/disperse crowds • Apply RCA to face and eyes for full effect	• Isolate area to prevent reinforcements • Prepare disposition plan for cleared personnel • Consider use of facility • Prepare security plan for cleared facility
Clear facilities of personnel	Logistics • Distribute NLW supplies • Resupply NLW munitions • Plan for transportation of displaced civilians • Plan facility maintenance or repair	Riot Control • Xenon searchlight • Flex cuffs • RCA markers • Flash bangs	• Searchlights dazzle/disorient individuals • Use flex cuffs for agitators • Use RCA markers to mark agitators for future ID • Flash bangs to dazzle/disorient individuals	
Clear facilities of personnel	Civil Affairs • Collect civilian intelligence • Care for injured civilians • Coordinate plans for controlling civilians			
Clear facilities of personnel	Medical • Hospital planning • Medic planning for tactical units • Procurement of medical supplies			

4. Staff Planning/Responsibilities

a. Administration. A staff composed of the G-1/S-1, PAO, SJA, and others as needed, is responsible for advising the commander on NLW employment and ROE. Additionally, administrative personnel are responsible for processing all civilian detainees and claims that arise from NLW employment. As a minimum, the staff must consider the following:

(1) SJA/claims planning.

(2) Chaplain.

(3) Casualty notification.

(4) Detainee disposition plan.

(5) PAO/media planning.

(6) Information dissemination.

b. Intelligence. The G-2/S-2 office is responsible for conducting the IPB process and providing information on the adversary's culture, political stability, and level of dedication to the cause, motivation, leadership, capabilities, equipment, and tactics. The G-2/S-2 should locate native language speakers to assist the interrogator-translator team.

(1) Background on country and culture.

(2) Route/map reconnaissance.

(3) Place sensors for advance notification.

(4) Information on leaders.

(5) LP/OP.

(6) Determine reason for the crowd.

(7) Identify threat.

(8) Analyze traffic.

c. Operations. The G-3/S-3 office is responsible for unit training, nonlethal certification of units, mission analysis, maneuver plans, and CA planning when required.

(1) Tactically integrate NLW and lethal fires.

(2) Fire support plans to include using RCA.

(3) Allocate NLW munitions.

(4) Plan engineer barrier.

d. Logistics. The G-4/S-4 office is responsible for logistics considerations, ship-to-shore movement of weapons and ammunition, storage and maintenance of equipment, resupply, transportation ashore, and other support requirements inherent in using NLW.

(1) Distribute NLW supplies.

(2) Security of rear area facilities.

(3) Resupply NLW munitions.

(4) Special maintenance requirements for NLW equipment and munitions.

e. Communications Officer. The G-6/S-6 staff is responsible for installing, operating, and maintaining communications from units confronting belligerents to the tactical operation center (TOC) or combat operations center (COC). The staff must plan for the effects NLW will have on communications to include proper frequencies.

f. PAO. The PAO provides liaison with outside media and participates in the staff planning process. PAO is also responsible for ensuring the NLW media plan and news releases consider the perceptions of the native media and belligerents or adversaries. PA personnel must receive training on NLW technologies, employment and capabilities, acute and long-term effects, impact on the environment, and legal approval procedures (at a minimum) so they can respond to media and public inquiries.

g. Medical. Medical personnel must be aware of the effects of NLW. Medical personnel should also be familiar with and able to provide decontamination of RCA as well as administer antidotes in cases of adverse reactions to riot control and other agents.

(1) MEDEVAC planning.

(2) Procuring medical supplies.

(3) Hospital planning.

(4) Medical planning for tactical units.

h. Chaplain. The commander should consider chaplain support for counseling and/or religious services for military and civilian personnel. Additionally, chaplains should provide information regarding the religious and cultural background of the local populace.

i. CA. The CA office is responsible for planning and coordinating all aspects of CA activities and providing assistance in gathering intelligence from civilian sources.

(1) Handle displaced civilian personnel.

(2) Coordinate plans for control of civilians.

j. PSYOP. Tactical PSYOP forces serve as an ideal tool for the commander to resolve issues of armed resistance and local interference with military operations while encouraging dialogue and cooperation with noncombatants. This force provides dissemination of PSYOP products (loudspeaker messages, leaflets, posters, etc.) and conducts face-to-face communications with target audience. PSYOP forces provide the supported commanders with nonlethal leverage by influencing foreign populations and expressing information subjectively to influence attitudes and behavior, thereby obtaining compliance or noninterference. They provide public information to foreign populations to support humanitarian activities, ease suffering, and restore or maintain civil order. They serve as the supported commander's voice to foreign populations to convey intent, countering enemy propaganda, misinformation, and opposing information to correctly portray friendly intent and actions. Finally, PSYOP forces deny others the ability to polarize public opinion and political will against the United States and its allies within an AO.

5. Tactics

The commander should include the following considerations in preparing for unit tactical operations involving NLW:

a. Avoid reconfiguring small-unit organization, except for special configurations such as—

(1) Special teams.

(2) Attachments.

(3) Crowd control formations.

(4) Security organizations.

b. Standardize postures in unit tactical SOP. Take the following steps to add a higher volume of specialty impact munitions and equipment within squads, platoons, or units:

(1) Avoid arming troops involved in apprehending designated personnel with long weapons (i.e., shotguns, rifles) because they can be easily grabbed/pulled away by individuals.

(2) Add weapons systems (shotguns) that provide a greater nonlethal capability. When loaded with nonlethal munitions, the longer reloading time of the M203 grenade launcher is protected by coverage from the high rate of fire from shotgun.

(3) Add nonlethal munitions to existing organic weapons systems (M203 grenade launcher).

(4) Avoid arming troops in the front lines of crowd control formations with long weapons.

(5) Focus riot-control capabilities in specialized reaction forces.

(6) Establish base line NLW postures.

(7) Establish command relations for special units.

(8) Maximize distance and barriers.

(9) Create NLW range cards for static positions.

(10) Provide the force with available OC/RCA consistent with ROE.

(11) Always mutually support forces utilizing NLW with lethal means.

(12) Consider environmental conditions (heat/cold/wind) and their effect on the performance of nonlethal munitions. Proper storage, periodic inspections, and rotation of NLW ammunition are critical to maintaining the effectiveness and viability of nonlethal munitions.

c. Unit Weapons, Unit Configurations, and Ammunition Considerations.

(1) Individual. Designate individuals as nonlethal shooters. Nonlethal shooters should also carry lethal munitions for use as required. Lethal rounds must be marked and carried in a manner to prevent confusing nonlethal rounds with lethal rounds.

(2) Squad. Squad leaders should dictate who carries nonlethal grenades and maintain their weapons loaded with lethal ammunition. Ideally, the squad should not change its task organization to accommodate nonlethal equipment, but should designate nonlethal shooters.

(3) Patrols. Commanders should not plan a nonlethal patrol, but should plan a combat/security patrol with a nonlethal capability when mission dictates.

(4) Static Positions. Individuals in static positions should have their weapons loaded with lethal munitions (mission, enemy, terrain and weather, troops and support available–time available and civil considerations dependent) (METT-T/METT-TC). Additionally, they should carry nonlethal munitions. Both shotgun and M203 work well from static positions and should complement each other in a nonlethal mode.

(5) Recovery Teams. Recovery teams have the primary responsibility of retrieving and controlling designated personnel from the crowd; the commander should establish a recovery team before deploying. The force should train team members in unarmed self-defense, open-hand control, and hand cuffing/flex cuffing. It should equip recovery teams with personal protection gear, 9-mm weapon, and flex cuffs. Due to the physical nature of recovery teams, it is important not to arm team members with long rifles. If security for the recovery team is an issue, the commander should attach a security element.

(6) Crowd Control Formations. The unit should establish crowd control formations within a minimum response time. Because of the physical nature of crowd control, individuals in these formations, if possible, should not carry long rifles. These formations should have inherent nonlethal capability. The unit must provide lethal coverage for the entire formation.

d. Individual Elements of a Formation. These elements are: base element, support element, and command element. The formation needs to provide versatility. Restrictions must never be set on numbers, equipment, or changes to the standard formations.

(1) Base Element. Role of the base element (front line):

(a) Base element serves as the front line or the base/foundation of the formation (can be divided into two or three sections or ranks).

(b) Front line consists of the shield holders.

(c) Angle the bottom of the shield to channel flammable liquids away from the body.

(d) Distance or interval between the shields differs based upon the range to the crowd. The closer the crowd, the tighter the interval between the shields.

(e) Provide frontal security.

(f) Hold a solid line.

(g) Provide intimidation factor.

(h) Fire team leaders are in the line holding shields.

(i) Secondary (tertiary) line(s) can consist of the baton holders and/or nonlethal gunners.

(j) Baton holders will—

- Strike with batons.
- Spray OC or disperse smoke.
- Serve as spotters for the NLW gunners.
- Control and detain subjects pulled from the crowd.

(k) Nonlethal gunners will—

- Consist of team leader with M203/M16 (provides nonlethal cover with M203 40-mm munitions and lethal cover with M16).
- Position themselves between shields working with baton holders.
- Serve as immediate control for the team.
- Maintain continuous muzzle awareness of the weapon.
- Provide immediate support to the shields.

(2) Support Element. Role and employment of the support element:

(a) Provide the means to employ lethal and nonlethal capabilities.

(b) Can consist of two ranks: M16/M203 (2nd rank) Mossberg 12-gauge shotguns (3rd rank).

(c) Provide security for search and recovery team.

(d) Consists of munitions teams and/or recovery teams.

(e) Munitions teams provide the means to employ lethal and nonlethal capabilities consisting of M16/M203 and Mossberg 12-gauge shotguns.

(f) Recovery teams provide ability to detain and search rioters downed by nonlethal munitions. Recovery teams will—

- Extract vehicles or personnel from the crowd who are in immediate danger.
- Detain and escort downed rioters to rear of the formation.
- Restrain and search detained individuals.
- Communicate with the base element.
- Spray or employ RCA as needed.
- Fill in the gaps in the front line as needed.
- Insert into a confined or congested area too large for the formation.
- Use a soft base line that capitalizes on the munitions fired to deploy forward of the shields in order to drive the crowd back, retrieve personnel/gear, and apprehend key riot personnel. Employed in pairs (1 x shield man/1 x shotgun).

NOTE: Refer to Marine Corps Security Force (MCSF) formation examples.

(3) Command Element. Role of the command element is to provide command and signal to the formation and consists of the platoon leader/flight commander/ platoon sergeant/flight sergeant, and squad leaders.

(a) Platoon Leader/Flight Commander, Platoon Sergeant/Flight Sergeant—

- Pass verbal commands.
- Maintain order within formations.
- Gauge and evaluate use of munitions and resupply considerations.

(b) Squad Leaders—

- Effectively control the squad.
- Pass information up the chain.
- Employ OC as directed.

(c) Designated Marksman (DM). During a nonlethal engagement, using DM provides confidence and safety to those facing a riot. The DM in an overwatch position scans the crowd, identifies threats and designated personnel for recovery, and fires lethal rounds if a lethal threat is presented. Additionally, DM are ideally suited for flank security and counter sniper operations.

6. Employment Objectives

a. Innovative employment methods of NLW and ingenuity take on critical significance. NLW are employed in concert with other nonlethal and lethal systems to create a synergistic effect.

b. The measures for countering many nonlethal options are usually apparent, quickly learned, and readily available. Because they are not intended to kill, nonlethal options "teach" an adversary what to avoid in the future. Small unit commanders must prepare to keep one step ahead of motivated belligerents.

c. Many nonlethal munitions have both maximum effective and minimum safe ranges. Individuals short of the minimum safe range may suffer severe injuries or death while the effects of most nonlethal munitions are greatly decreased at longer ranges. To be effective, an adversary must be engaged within the optimum range of the munition. Figure V-1 illustrates the ranges for a variety of nonlethal munitions and systems.

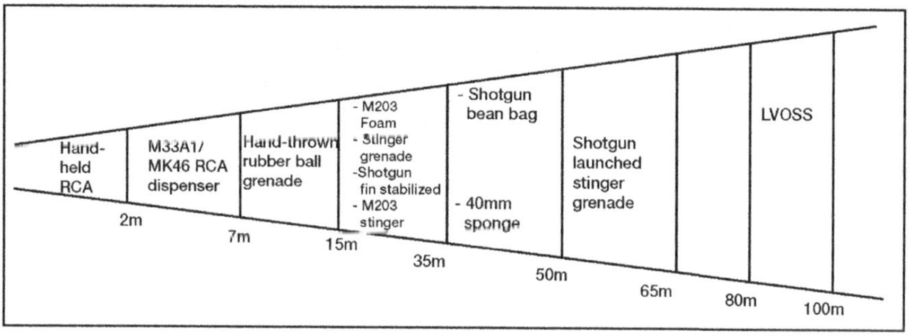

Figure V-1. Nonlethal Weapons Ranges

7. Rules of Engagement

a. U.S. forces follow directives issued by competent military authority that delineate the circumstances and limitations under which deadly force will be initiated against other forces. Intrinsic to the effective use of nonlethal options is the authorization to employ them. ROE are the primary means by which the chain of command can provide guidance to deployed forces in peacetime for handling crises and, in wartime, to help control the level of hostilities.

b. The commander should distinguish ROE from tactical control measures, threat conditions, and arming orders/weapons conditions. Although arming orders are sometimes listed on the same quick reference card as ROE for easy reference, the commander should distinguish ROE from other policies and directives developed, such as weapons confiscation rules.

c. The commander should delegate the decision to use NLW against an adversary to the lowest possible level, preferably to the platoon, flight, or squad level. However, this requires that all personnel, not just leaders, have a clear understanding of ROE and his intent.

d. Using lethal force, employed under the standing ROE, will never be denied. At no time will forces be deployed without the ability to defend themselves against a lethal threat nor will they forego normal training, arming, and equipping for combat. Nonlethal options are a complement to, not a replacement for, lethal force and seek to expand a proactive response across the range of military operations.

e. The following figures are examples of service-specific formations. These examples are not all-inclusive, and the tactical situation will always dictate actual unit employment.

FM 19-15, *Civil Disturbance*; MCWP 3-11.1, *Rifle Company;* and MCWP 3-11.2, *Marine Rifle Squad* provide extensive detail about small-unit formations, their use, and other considerations. Figures V-2 to V-7 are from I and II Marine Expeditionary Force (MEF) Special Operations Training Groups (SOTG) and represent a notional infantry platoon. Figures V-8 to V-13 represent a notional U.S. Army infantry platoon. Figures V-14 to V-21 are additional platoon formations submitted by Marine Corps Security Force Training Company.

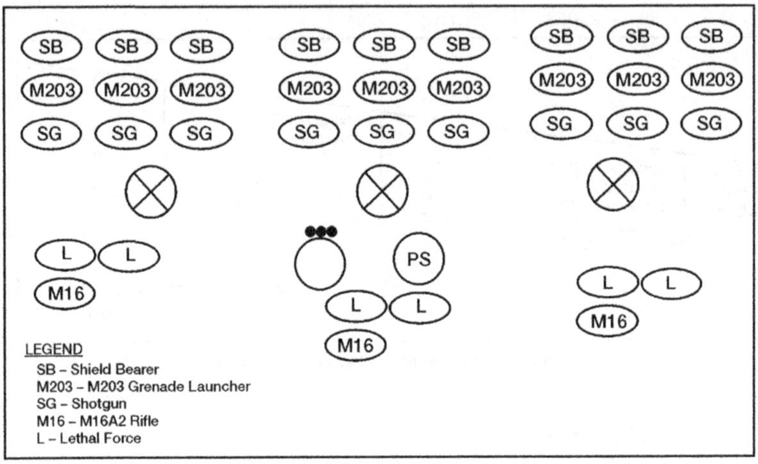

Figure V-2. Platoon in Depth

NOTE: The platoon is the smallest unit capable of independent maneuver. The platoon in depth formation shown above provides the following benefits:

- In-depth support with three ranks.
- Complementary support with the shotguns covering the slower reload time for the M203.
- Additional personnel from each squad to form the recovery teams.
- Each squad is capable of moving as an independent formation forward of the platoon line, if situation dictates.
- Other tactical formations are applicable—echelon, wedge, column.

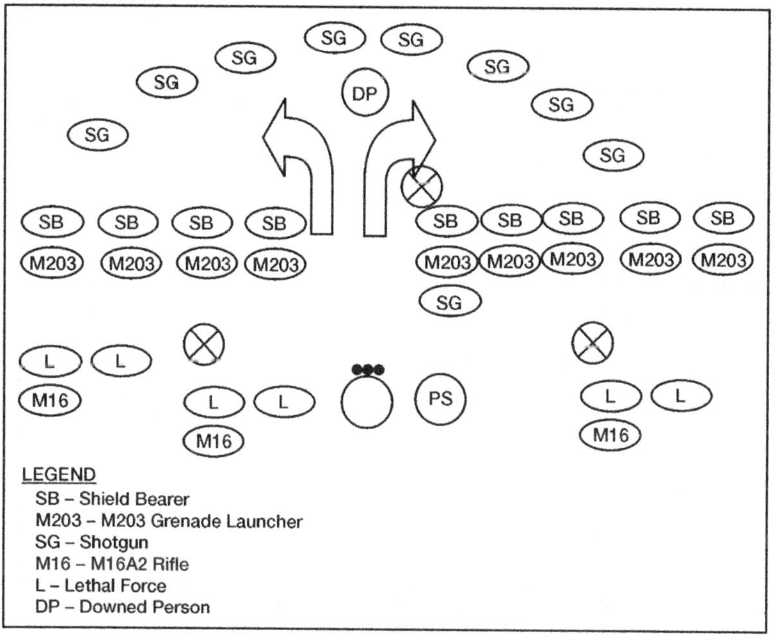

Figure V-3. Platoon Recovery Team Deployment—Security Element

---

NOTE: Security element deploys around downed person.

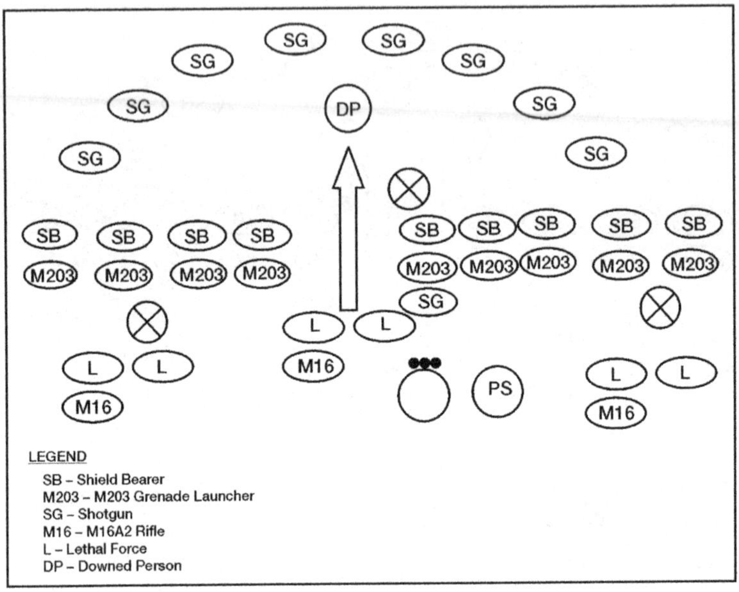

Figure V-4. Platoon Recovery Team Deployment—Recovery Team

NOTE: Recovery team deploys to detain, search, and recover downed person.

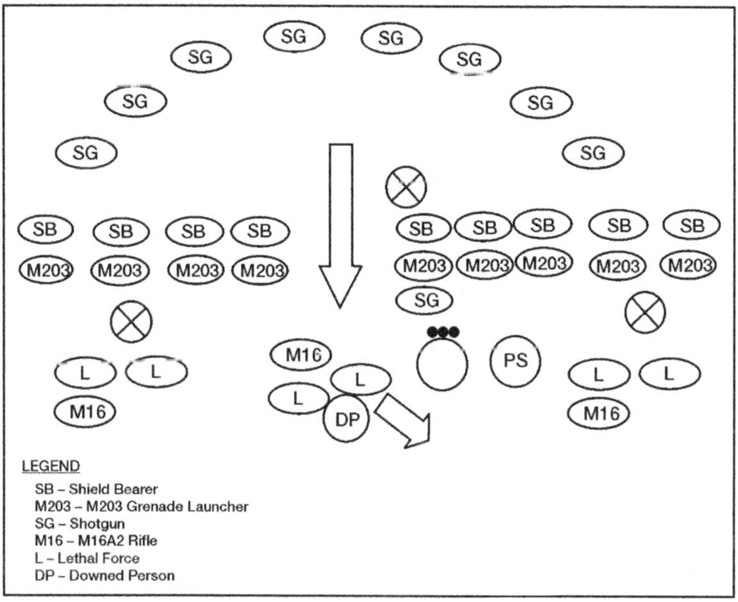

Figure V-5. Platoon Recovery Team Deployment—Recovery Team Redeploy

---

NOTE: Recovery team redeploys with detained person.

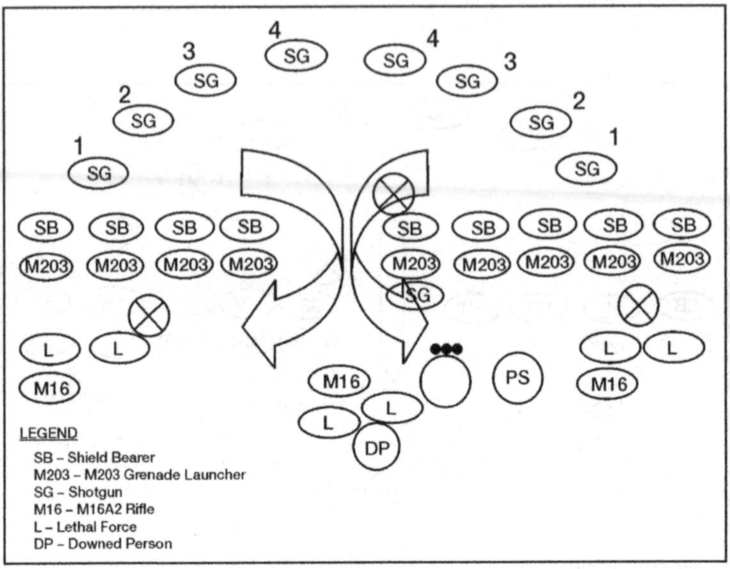

Figure V-6. Platoon Recovery Team Deployment—Security Element Redeploy

---

NOTE: Security element redeploys back to 3rd rank "by the numbers." Team leader calls "4s in," then "3s in," etc. The last person in calls "last man in" and formation closes the gap.

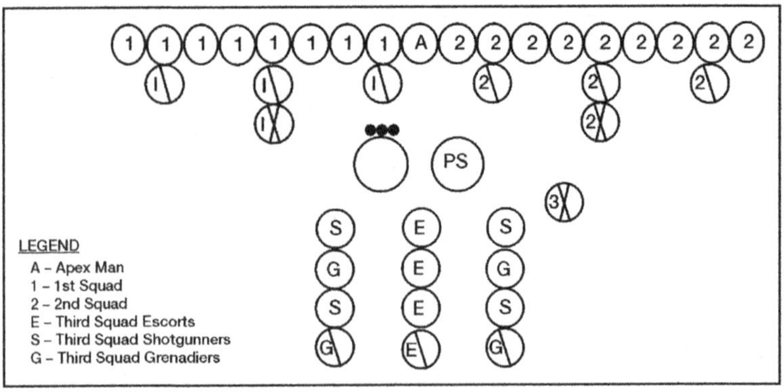

Figure V-7. Platoon On Line—Munitions Squad in Close Support to Cover Flanks

---

NOTE: Platoon in close support. NLW and lethal weapons are sited to cover the flanks of the formation, as well as support the front line.

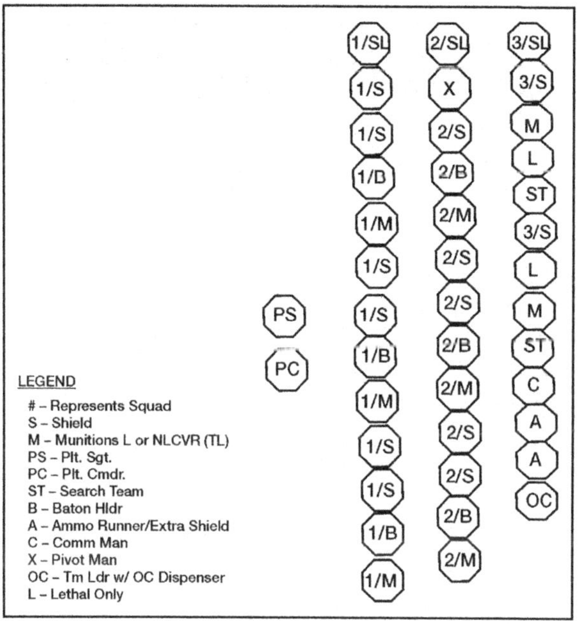

Figure V-8. Platoon Column

NOTE: Platoon in column formation. For administrative movement only.

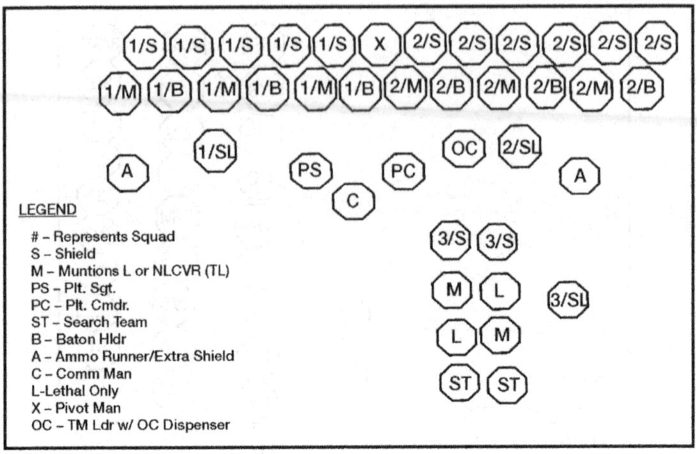

Figure V-9. Platoon On Line

NOTE: Platoon on line—basic formation for the employment of NLW.

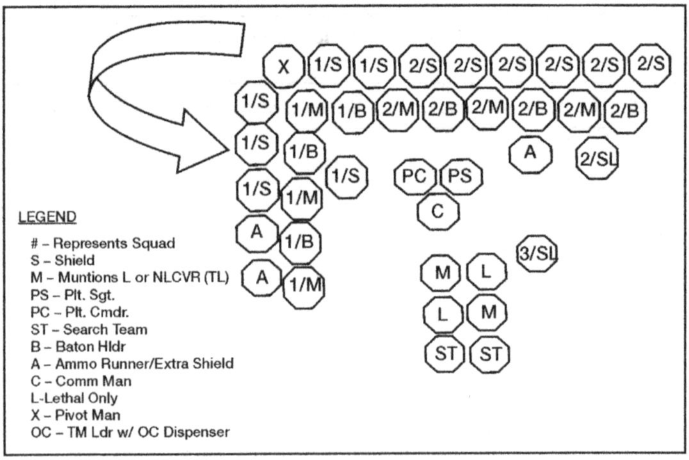

Figure V-10. Platoon Lateral Support, Left (Right)

NOTE: Platoon in lateral support to protect flanks.

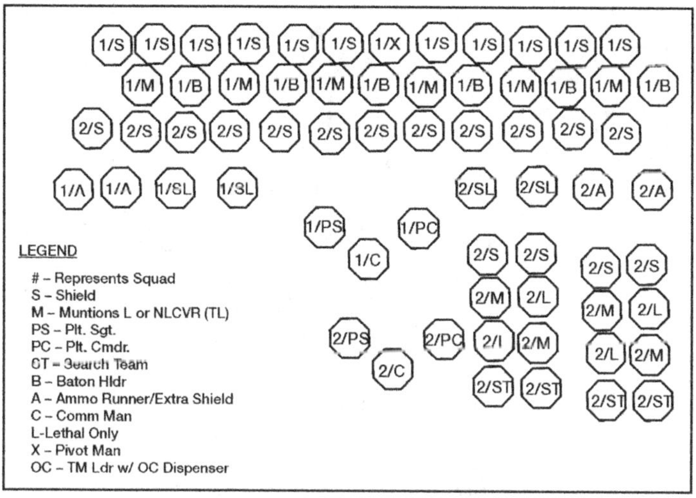

Figure V-11. Two Platoons in Close Support

NOTE: Two platoons in close support—tied in for mutual advantage.

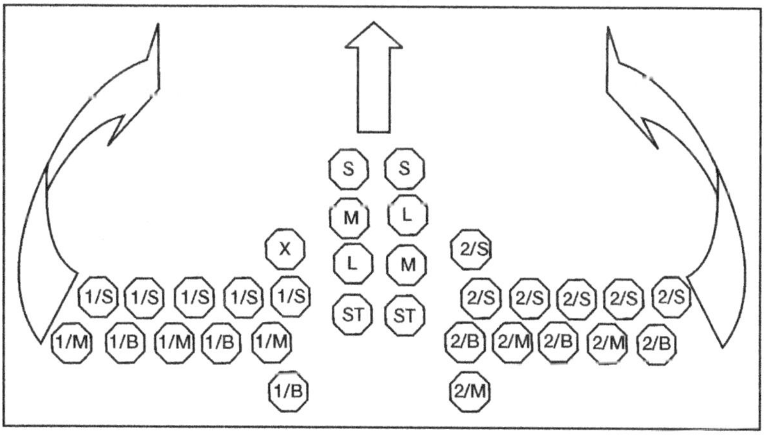

Figure V-12. Recovery Team Deployment

NOTE: Recovery team deploys from formation—remainder of platoon can deploy around the recovery team from either flank.

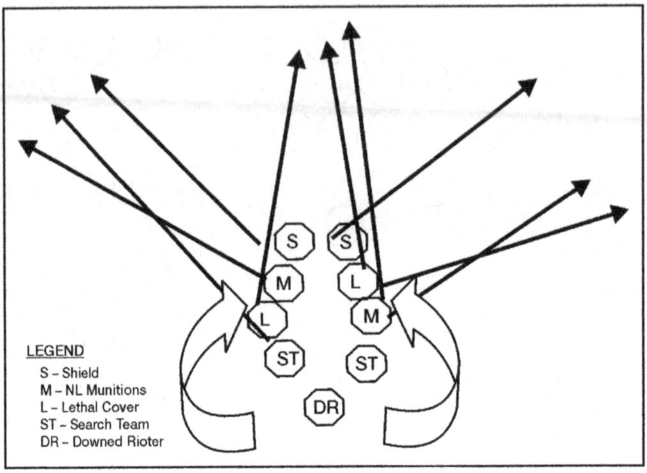

Figure V-13. Recovery Team Actions and Security Posture

NOTE: Recovery team forms security perimeter around downed rioter, conducts search and detain actions. Once actions are complete, recovery team redeploys.

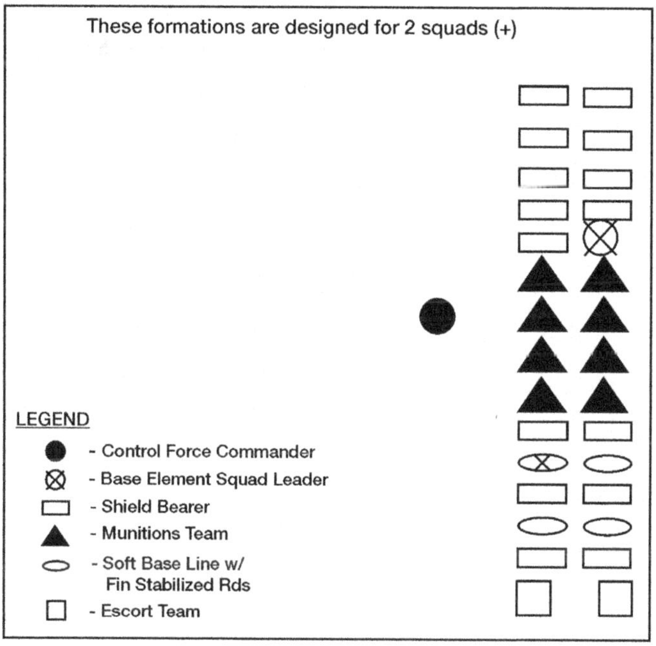

Figure V-14. Platoon in Column (USMC)

NOTE: These formations are designed for 2 squads (+). This allows the platoon to move rapidly when there is no chance of being engaged with the crowd.

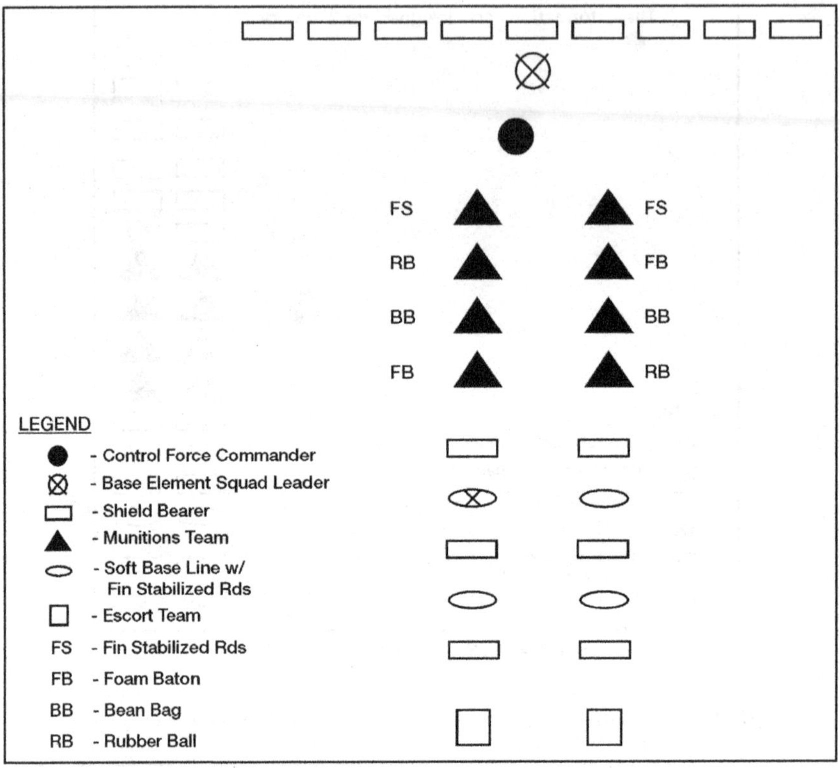

Figure V-15. Platoon in General Support (USMC)

NOTE: The control force commander moves into this formation approximately 60 to 70 meters from the crowd, as the crowd could start to engage the control force with thrown objects.

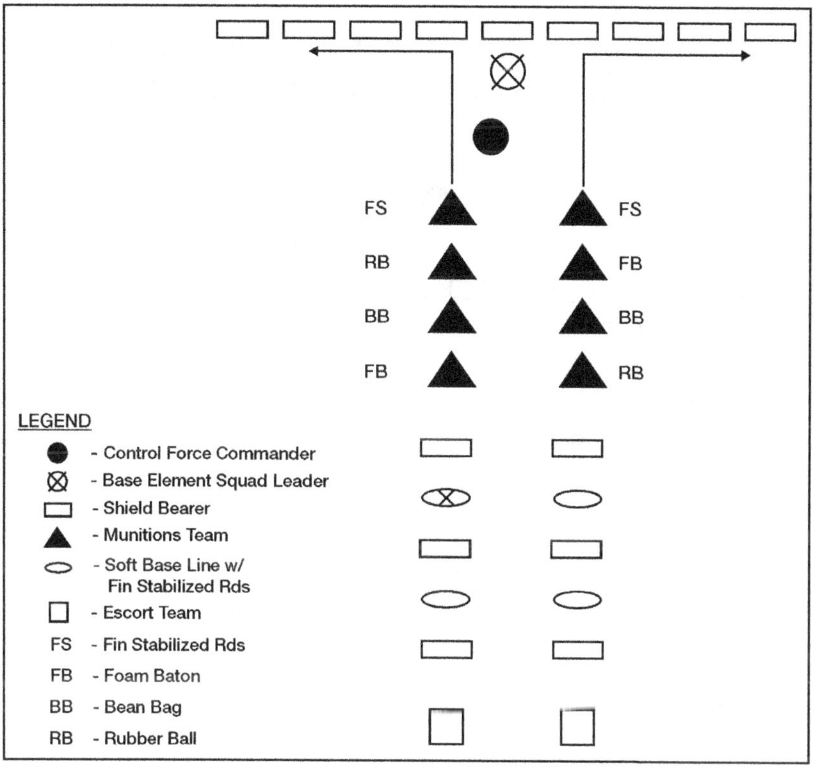

Figure V-16. Platoon in Close Support (Start) (USMC)

NOTE: The control force commander moves into this formation when he feels ready to engage the crowd with nonlethal munitions.

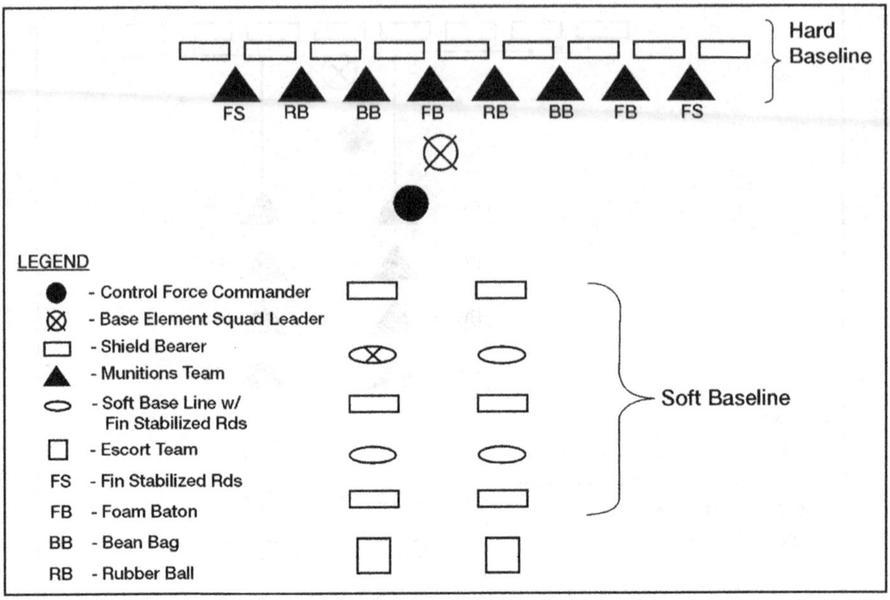

Figure V-17. Platoon in Close Support (Finish)

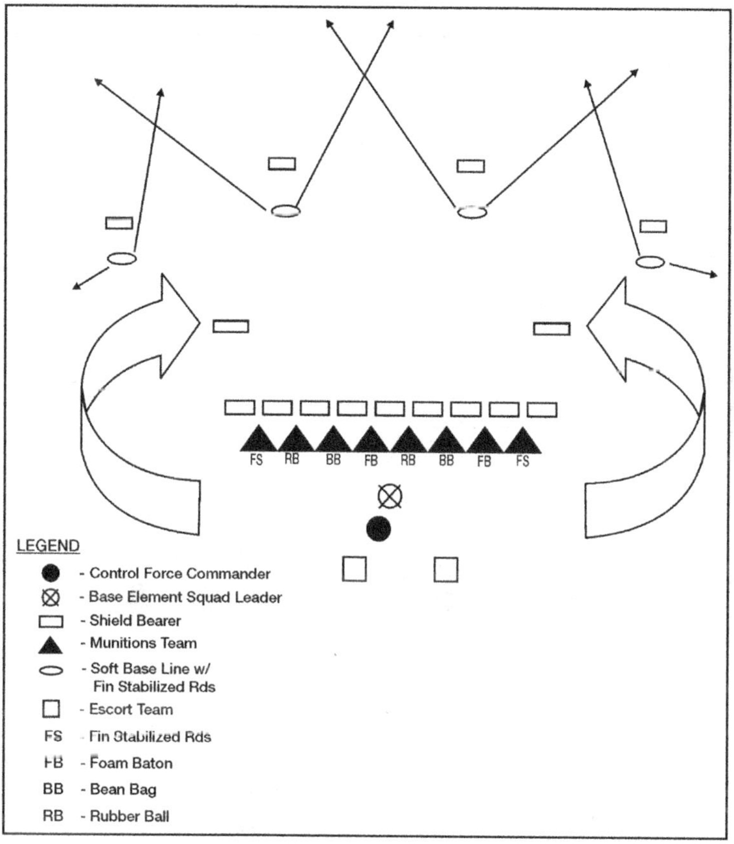

Figure V-18. Deploying Soft Baseline

NOTE: The soft baseline affords the control force commander much flexibility. The primary mission of this formation is to disperse the crowd back using speed and extreme violence of action. By deploying the soft baseline, the commander is able to maneuver the hard baseline where needed (around the barrier plan or obstacles.) The soft baseline is deployed by a predetermined signal (bullhorn siren, whistle blasts, vehicle horn) and will be retrieved using the same signal. This signal conditions the crowd to expect violence of action and causes them to disperse. The soft baseline should never be deployed into a fresh crowd; the crowd needs to be distracted by nonlethal munitions. The formula found to work is distract with stingball grenades, followed by a volley of shotguns and M203 munitions, then send the soft baseline out to disperse or move the crowd back. (Sting them, shoot them, soften them up.)

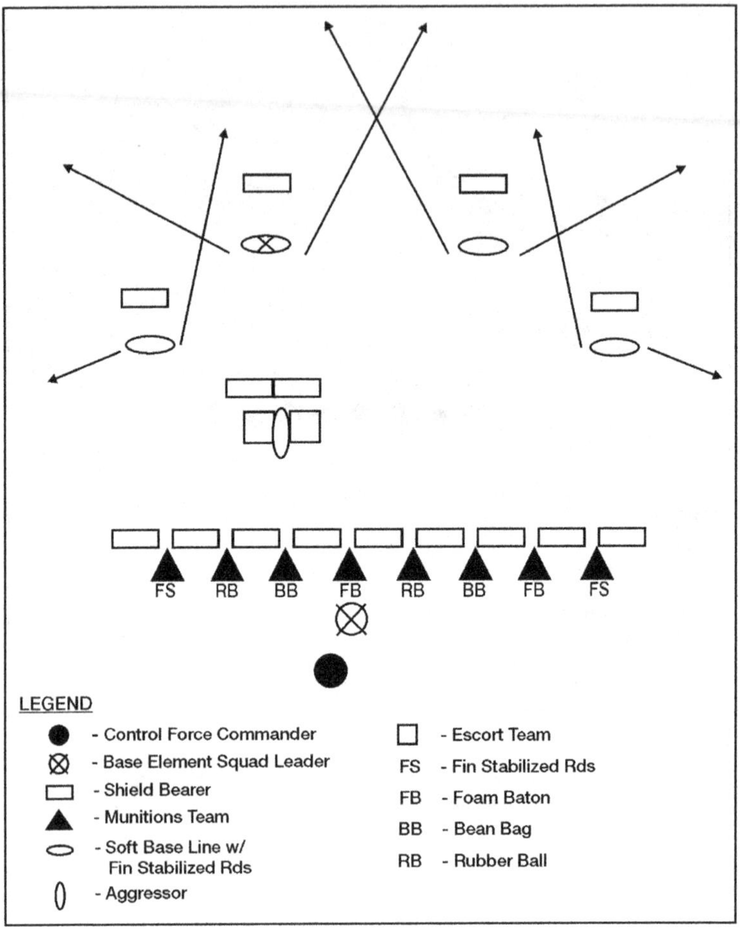

Figure V-19. Deploying Escort Team (Start)

NOTE: When the control force commander identifies the aggressor to retrieve, the command is, "Escort team left/right," then the signal. The signal is the same to deploy the soft baseline. As the soft baseline deploys, the two shields without gunners provide protection for the escort team. The shields remain with the escort team until the soft baseline is called back in, at which time the shields go back to the soft baseline.

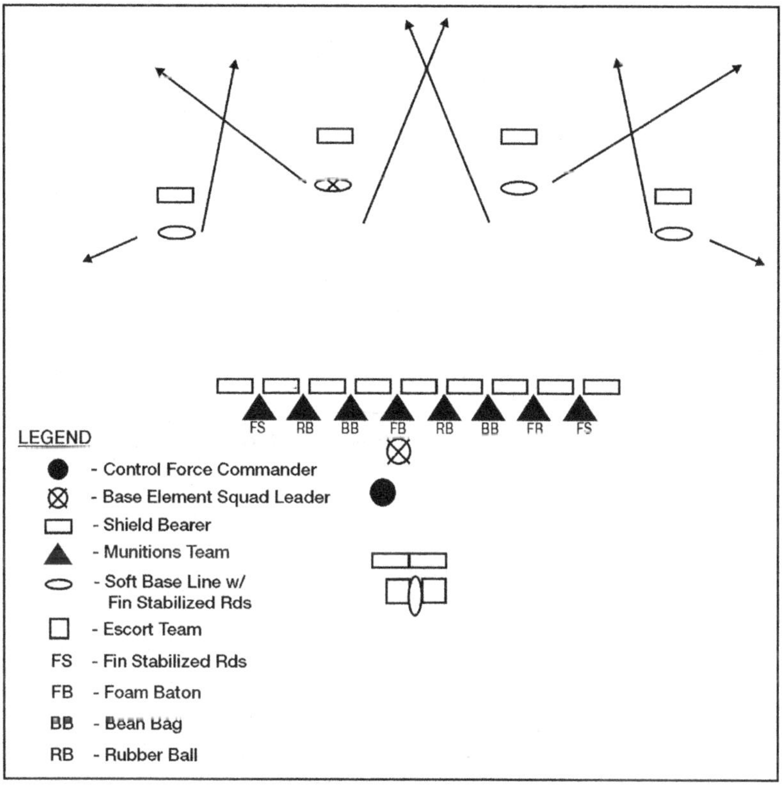

Figure V-20. Deploying Escort Team (Finish)

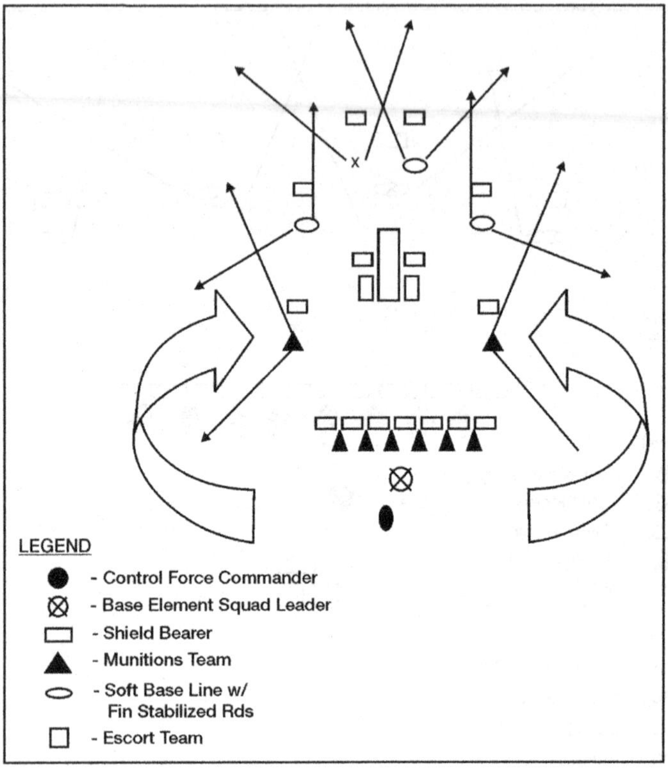

Figure V-21. Recovery of Downed Personnel or Stranded Vehicle (Big Johnson)

***NOTE*:** In the event the crowd attacks personnel or a vehicle, the control force commander can call for a "Big Johnson." In a Big Johnson, the soft baseline deploys as usual except now the shield and shotgunners on the flank of the hard baseline also deploy creating a 190-degree perimeter around the personnel or vehicle. This maneuver must be executed quickly and with extreme violence of action because the crowd already has its hands on individuals in the force. The escort team deploys with the Big Johnson in order to escort or carry injured individuals. The shields without gunners provide protection for the escort team.

Chapter VI

LESSONS LEARNED

1. Background

Although lessons learned from employing NLW are still being studied and impacts analyzed, it is beneficial to look at the preliminary lessons gained from the NLW experience. These lessons are far from conclusive, because U.S. forces have yet to see the first large-scale employment of NLW. However, U.S. forces can gain tremendous insight from the experience of others and use what has been learned to guide future NLW employment.

2. Lessons Learned

a. Force Continuum. Using nonlethal options needs to be based on the principle of a graduated and proportional response, using the decisive force necessary to respond to the threat.

b. Attempts to legislate and regulate every aspect of NLW employment can prove disadvantageous to the on-scene commander. During the evacuation of U.N. forces in Somalia (OPERATION UNITED SHIELD, 1995), the ROE made an arbitrary and unnecessary distinction between situations justifying deadly force and those that justified only lesser means. Crowd control items were restricted to situations requiring deadly force. Many soldiers, airmen, and Marines asked, "why resort to nonlethal means at all if I have to wait until deadly force is authorized before I can shoot them with a beanbag?" Additionally, the use of sticky foam was limited to deadly force situations because of the perception that adversaries might suffocate from its application. This stipulation effectively eliminated sticky foam as a nonlethal option to the forces in Somalia.

c. With ROE that take into account the force continuum and encourage graduated response to nonlethal threats, the commander can employ NLW to enhance his flexibility, deterrence ability, and reaction to uncertain situations. To accomplish the mission with minimum loss of life and property, commanders need to have ROE that do not restrict them from properly employing NLW.

Lessons Learned from OPERATION JOINT GUARDIAN 1999 (Battalion Landing Team (BLT) 3/8 26 Marine Expeditionary Unit (MEU) Peace Operations in Kosovo:

ROE significantly impaired the employment of NLW. For example, OC was not authorized because it was considered a chemical agent. Nonlethal direct fire munitions were not authorized during the initial occupation of the province. This significantly hampered the individual Marine's ability to neutralize a threat where deadly force would not have been justified but action was required. Overly restrictive ROE thus negated the flexibility of response that NLW are supposed to provide. The intent of deploying NLW is to provide the operator the capability and flexibility to neutralize a threat without having to apply deadly force. Thus, the ROE for NLW should ideally be less restrictive, not the same or more restrictive, than those for lethal force.

It is often enticing for authors of ROE involving the use of NLW to base those ROE on the principle of minimum force necessary. This may be appropriate for base and station or homeland security ROE; however, for expeditionary operations the decision to mandate minimum force deserves critical consideration. Commanders in military operations have traditionally gained advantage over their enemy not only by the actual combat power that they bring to the fight, but also from that enemy's perceptions and beliefs about the intensity and resolve with which that combat power will be applied. When ROE are written stating that the confrontations will be addressed using the minimum force necessary, this advantage is lost. ROE developed for expeditionary nonlethal operations should be based on the concept of decisive and compelling force, rather than the minimum force. This will maintain an adversary's hesitancy to harass and endanger U.S. forces even after that adversary has tested or probed the response of those U.S. forces. Further, this will allow synergy to be gained through the complimentary employment of lethal and NLW by providing more response options to the small unit leader.

3. Employment

a. Through the use of modeling, simulation, and field exercises, new concepts for employing NLW are continually being developed. The force should train with NLW to the same degree as for lethal weapons. In this way, units become confident when using nonlethal options in chaotic and uncertain situations. The commander should push the decision to employ NLW to the lowest possible tactical level. Commanders who intend to use NLW in achieving military objectives must recognize the fluidity of the situations in which they are likely to be employed. The decisionmaking cycle is short, stressful, and the level of violence can move in both directions along the force continuum. In quick succession, a small unit leader might employ nonlethal munitions, decide to employ deadly force to an immediate specific threat that arises, then return to using nonlethal means in response to unarmed hostile elements.

b. A mixture of nonlethal "tools" has been determined to be advantageous in an operational scenario. As the situation develops, commanders can apply different nonlethal responses with respect to range/distance capability and severity of effect. For example, to disperse a crowd, area fire weapons would precede use of close-in point target nonlethal munitions.

c. An important point to remember is that NLW are just one element of a graduated response available to the force commander. Other factors make up a large contribution to the success of U.S. forces, including barrier plans and PSYOP. An example of this operation would be using barriers such as roadblocks or concertina wire, warning commands to inform a mob of potential responses based on their actions, and only then taking actions through escalated employment of NLW. These continued responses would involve smoke, RCA, and blunt munitions. A specific verbal warning should precede every escalation of nonlethal means.

d. Whenever NLW are employed, a separate lethal capability must always be in place. Troops must always be prepared to protect themselves as required by lethal means, because it is impossible to accurately predict the response of an individual or mob to NLW employment.

Lessons Learned from OPERATION ESSENTIAL HARVEST 2001 (U.S. Embassy Reinforcement operations conducted at the U.S. Embassy Skopje Macedonia by 5th Platoon, 2nd Fleet Antiterrorism Security Team [FAST] USMC):

The anticipated threats were violent demonstrations directed against U.S. facilities in the AO. In this case, USMC forces tasked to reinforce the embassy were equipped with a compliment of NLW. ROE took into account the capabilities and limitations of the NLW. The ROE and the NLW systems on the ground were coordinated and thus compatible. In addition, it was found that integrated employment of different nonlethal munitions greatly enhanced that force's ability to suppress the threats encountered (large rioting mobs). Random security patrols reinforced with military working dogs proved very effective. Reinforcement with multiple M203s provided USMC units with both a lethal and nonlethal capability in the same weapons system. The M203 was particularly useful as it gave the individual Marine the means to quickly transition between a lethal (5.56-mm) and nonlethal (40-mm nonlethal) capability. A basic understanding of the capabilities and limitations of nonlethal munitions enabled a relatively small force (40-50 Marines) to counter a sizable threat many times its size (500-1000 rioters). Although NLW were not used, the capability and preparation of the units provided the adequate deterrence that prevented the situation form escalating.

4. Publicity and Media

a. The commander can use the media to his advantage, as they will inform the local population of the military's capabilities and responses available. During OPERATION UNITED SHIELD in Somalia, the local media was informed of the Marines' nonlethal options and "new technologies" in order to discourage hostile actions. The media was encouraged to report this to the population. Intelligence had indicated the Somalis were planning to mix civilians with armed clansmen, creating hostile mobs. In fact, feuding warlords had previously successfully demonstrated this strategy on several occasions. Lack of deadly mob incidents implies Somalis were not anxious to test this new technology. Evidently, publicity had directly contributed to mission accomplishment.

b. The public dissemination of the presence of NLW capabilities must strike a balance between information sufficient to discourage hostile actions and information that enables the local population to learn too much of the capabilities of NLW. Too much information allows adversaries to develop countermeasures or avoid the NLW effects. Commanders who employ NLW systems must involve the PAO from the beginning of planning. The PAO must be familiar with the objectives of the mission, commander's intent, ROE, NLW and lethal weapons employed, and be prepared to advise the commander on the best methods of dealing with outside media. The commander should fully integrate the PA plan with PSYOP objectives and the concept of operations. Additionally, he must work the media to ensure the "sensationalism" of such a new technology does not create unrealistic expectations where he can be unfairly judged or constrained.

5. Capabilities and Impact on Mission Accomplishment

a. The commander should understand the unique capabilities of NLW and use them to accomplish the mission. He should use NLW in conjunction with the planned employment of the force and within established ROE to minimize loss of life and property while accomplishing the mission. The force requires early hands-on training to recognize the capabilities and limitations of NLW. Appendix D has sample training plans developed to meet these objectives.

b. The force commander must fully integrate planning for using nonlethal capabilities as quickly as possible in the staff planning cycle. In order to have a full range of alternatives for use against unarmed but hostile elements, the force needs to deliver nonlethal systems, complete necessary ROE and legal review, and make sufficient quantities of each NLW readily accessible to the force. The choice of nonlethal capabilities should be based on several factors and the choice can be made after addressing these questions:

(1) First, are the NLW systems available and can they be delivered on time?

(2) Second, are they compatible with, and do they complement existing weapons and training?

(3) Third, are there clear and unambiguous guidelines and ROE for employment?

(4) Finally, can the NLW be effectively employed to save lives and contribute to mission accomplishment?

c. Leaders need to anticipate local civilians developing effective countermeasures to their NLW and tactics. Therefore, prior to beginning their mission, leaders must determine and rehearse actions to take when nonlethal munitions are effectively countered.

d. Using nonlethal and lethal force are not mutually exclusive. In many instances, the commander can gain synergistic effects by using a combination of both. The following are some examples of this:

(1) Utilizing NLW to separate agitators from noncombatants with simultaneous employment of lethal force against hostiles or combatants where deadly force is justified.

(2) Utilizing NLW when making entry into structures or rooms to gain time in order to discern temper and intent of occupants prior to applying deadly force.

6. Knowing Your Adversary

a. Although commanders desire thorough intelligence of any potential adversary, such a goal may be difficult to reach in a NLW employment environment. During OPERATION UNITED SHIELD in Somalia and OPERATION UPHOLD DEMOCRACY (1995) in Haiti, a study of the local population's culture, history, and customs enabled commanders to anticipate and control many situations before they got out of hand. On the other hand, during OPERATION UPHOLD DEMOCRACY, poor intelligence and a lack of understanding between the two cultures led to the Haitians and Cubans being lodged together in Guantanamo Bay. This situation led to riots between factions who had a deep dislike for each other. By understanding a particular crowd's motivation, friendly forces may be better able to take necessary steps to prevent escalation resulting in the use of deadly force. Early training in crowd dynamics and riot control will enhance the effective use of NLW.

Lessons Learned from U.S. Peace Operations, 739th MP Battalion (Bn), January 2001

An Army MP platoon received a mission to clear a rioting crowd of 300 civilian locals from a main supply route. The platoon placed vehicle-mounted machine guns in overwatch positions to allow quick transition from NLW to conventional rounds. 12-gauge shotgun teams armed with fin stabilized rounds carried M9 pistols for lethal backup. Two volleys of shotgun and M203 nonlethal rounds successfully dispersed the crowd. The platoon had utilized INIWIC graduates to conduct preparatory training on using NLW. The NLW were extremely effective in pushing back the crowd and causing it to disperse. The 40-mm rubber ball round was moderately effective at close range, but very inaccurate at longer ranges. Shotguns were more effective than the M203 due to their high rate of fire and quick-loading capabilities.

b. Specific lessons learned during this operation are as follows:

(1) Placing interpreters is critical. The solution in this incident was to place interpreters approximately 25 meters behind formation so they could deploy forward as needed. Interpreters were often needed to communicate with crowds and with adjacent forces and coalition units that were organic to Kosovo forces (KFOR) that also responded to the incident.

(2) Leaders as well as soldiers should prepare to speak to the media. The command should have a PAO representative on hand to deal with the media.

(3) The commander should thoroughly brief the transition to lethal force from NLW.

(4) U.S. commanders in Kosovo engaged in the decisionmaking process had disparate levels of knowledge of nonlethal capabilities, constraints, and operational considerations. Therefore, leader training and orientation requires more emphasis.

(5) ROE were vague regarding the use of NLW. In many operations, NLW use is delegated to the on-site commander. Delegating to the lowest possible level is the preferred method.

(6) A unit needs a mixture of nonlethal rounds based on the size of crowd, composition of the crowd, available weapons systems, and synergistic effect of selected nonlethal combinations.

(7) When a RRF/QRF arrives on station with NLW to support a force equipped only with lethal means, the commander should address the following command and control concerns:

(a) Who makes the decision to employ NLW?

(b) Is the reinforced commander familiar with NLW?

(c) Who controls the reinforced unit?

(d) How does this relationship change as the situation develops?

7. Summary

a. The wide range of military operations in which U.S. forces are involved has demanded the introduction of NLW into the arsenal of weaponry available to commanders. NLW provide a graduated and proportional response to threats posed by a variety of adversaries. The ROE must be consistent with the rapidly changing situation of today's chaotic and uncertain military operations. This reality demands a high level of training in properly employing NLW.

b. Effective use of NLW involves a synergy between PSYOP activities, intelligence gathering, and professional media/PA plan. NLW effectiveness is enhanced through integration of these activities to reduce the chance adversaries will be able to develop effective countermeasures. NLW, when combined with lethal weapons, provide the commander with a force response across a wide range of military operations.

Appendix A

DOD Nonlethal Weapons and Equipment Item Description

The following is a cumulative list of NLW and equipment currently fielded within the DOD. Each service has various components of the following list in its inventory. A list of the service's items is located in Appendix E.

 **Figure A-1. Riot Face Shield**	1. Riot Face Shield: The riot face shield is a non-ballistic face shield designed to protect the service member's face from objects thrown or attacks from non-ballistic weapons. (Figure A-1)
No picture available.	2. Ballistic Face Shield: The ballistic face shield provides the service member with face protection from a wide range of threats (i.e. debris, liquids, handthrown objects). The shield is manufactured from acrylic and bullet resistant materials to provide Level IIIA protection (9-mm and .44 magnum pistol rounds), weighs 3.4 pounds and is compatible with the PASGT helmet and M-17/M-40 protective masks.
 **Figure A-2. Riot Body Shield**	3. Riot Body Shield: The riot body shield is a non-ballistic body shield designed to protect the service member's body from objects thrown or attacks from non-ballistic weapons. (Figure A-2)
No picture available.	4. Ballistic Body Shield with Light Kit: The ballistic shield provides the service member with torso protection from a wide range of threats (i.e. debris, liquids, hand thrown objects). The shield is manufactured from Spectra Shield resistant material painted anti-reflective flat black, and is capable of providing Level IIIA protection (9-mm and .44 magnum pistol rounds). The shield weighs 18 pounds and is 20 x 36 inches in size with a view port of 4 x 16 inches.

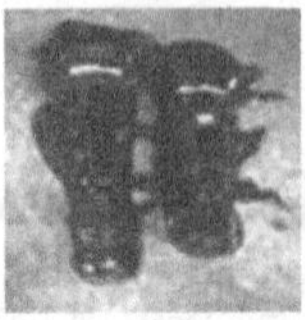 Figure A-3. Riot Shin Guard	5. Riot Shin Guard: The riot shin guard is a non-ballistic shin guard designed to protect the service member's shins from objects thrown or attacks from non-ballistic weapons. (Figure A-3)
No picture available.	6. Ballistic Shin Guards: Ballistic shin guards provide the service member with leg protection from a wide range of threats (i.e. debris, liquids, hand thrown objects). The guards are manufactured from Kevlar KM2 and provide Level IIIA protection (9-mm and .44 magnum pistol rounds). They are available in three sizes and weigh between 7 to 10 pounds.
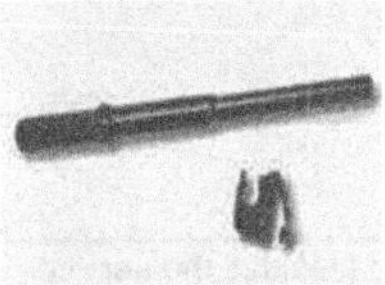 Figure A-4. Expandable Riot Baton (with holster)	7. Expandable Riot Baton (with holster): The expandable riot baton provides the service member a nonlethal means of crowd control and self-defense. The baton comes with a mounting device that attaches the baton to a belt. The baton is 24 to 36 inches in length. (Figure A-4)
No picture available.	8. Wooden Baton: The wooden baton provides self-defense for the service member and the option between physical force and lethal force. The baton comes with a mounting device that attaches the baton to a belt. The baton is 36 inches in length.
No picture available.	9. Practice Riot Batons: The practice riot batons provide the service member with a safe dynamic training environment for properly using a riot baton. The batons are available in a variety of lengths.
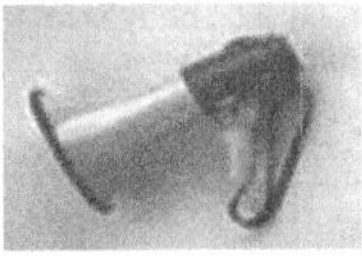 Figure A-5. Portable Bullhorn	10. Portable Bullhorn: The portable bullhorn provides the service member the capability to control his forces by projecting his voice above the noise and commotion created by crowds and mobs. (Figure A-5)
No picture available	11. Ground-Mounted Public Address System: The ground-mounted public address system is a communication enhancement device for conducting crowd control tactics. The system facilitates communication with the crowd in the absence of PSYOP support and assists with communication of commands to troops engaged in crowd-over-crowd noise.
No picture available.	12. M7 Individual Voice Amplification System: The M7 individual voice amplification system is a communication enhancement device for conducting crowd control tactics while wearing the M40 protective mask and using riot control agents. The M7 is fitted to the M40 protective mask to facilitate oral communication and increase the user's ability to communicate using radios and other devices.

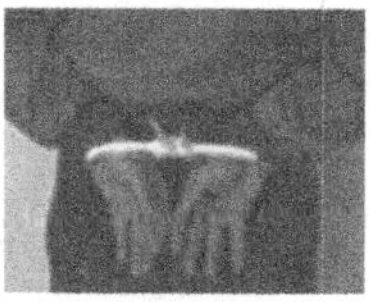

Figure A-6. Disposable Handcuff

13. Disposable Handcuff: The disposable handcuff provides service members with a means of restraining and controlling suspects, prisoners, and other individuals encountered during crowd control. The disposable handcuff is designed for one-time use. (Figure A-6)

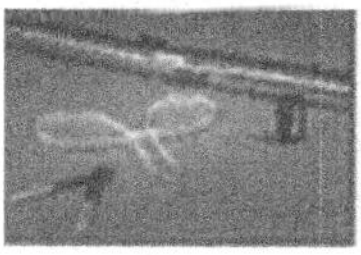

Figure A-7. Training Disposable Handcuff

14. Training Disposable Handcuff: The training disposable handcuff provides service members with a means for training in handcuffing procedures prior to actual use. The training disposable handcuff is red and reusable. (Figure A-7)

Figure A-8. MK-4 Pepper Spray Pouch

15. MK-4 Pepper Spray Pouch: The MK-4 pepper spray pouch is designed to carry the MK-4 pepper spray canister. (Figure A-8)

Figure A-9. MK-4 Individual Live OC Pepper Spray

16. MK-4 Individual Live OC Pepper Spray: The MK-4 individual pepper spray has a range of up to 12 feet in a target-specific stream. It contains 30 to 35 half-second bursts. (Figure A-9)

No picture available

17. M-36 Individual Riot Control Disperser: M-36 Individual Riot Control Disperser has a range of up to 12 feet in a target-specific stream. It contains 30 to 35 half-second bursts. It is loaded with either CN or CS agent.

Figure A-10. MK-9 Pepper Spray Pouch

18. MK-9 Pepper Spray Pouch: The MK-9 pepper spray pouch is designed to carry the MK-9 pepper spray canister. (Figure A-10)

Figure A-11. MK-9 Squad Live OC Pepper Spray

19. MK-9 Squad Live OC Pepper Spray: The MK-9 squad pepper spray has a range of up to 15 feet in a target-specific fog. It contains 8 to10 half-second high volume bursts. (Figure A-11)

Figure A-12. MK-46 Platoon Live (Horizontal) OC Pepper Spray

20. MK-46 Platoon Live (Horizontal) OC Pepper Spray: The MK-46 platoon pepper spray has a range of up to 25 to 30 feet in a target-specific dispersed spray pattern. It contains up to 12 one-second high volume bursts. (Figure A-12)

Figure A-13. MK-46 Sling

21. MK-46 Sling: The MK-46 sling gives a service member the ability to sling the MK-46 canister on the shoulder. (Figure A-13)

Figure A-14. MK-46 Hardware Kit

22. MK-46 Hardware Kit: The MK-46 hardware kit consists of a transfer tube, funnel, and transfer bottle. A compressed nitrogen source is also required to utilize the kit. (Figure A-14)

Figure A-15. MK-46 Refill Live OC Pepper Spray

23. MK-46 Refill Live OC Pepper Spray: The MK-46 refill solution comes in a one-gallon container. (Figure A-15)

Figure A-16. MK-4 Individual Inert OC Pepper Spray

24. MK-4 Individual Inert OC Pepper Spray: The MK-4 inert individual pepper spray has a range of up to 12 feet in a target-specific stream. It contains 30 to 35 half-second bursts. It provides the service member with an inert canister for familiarization and training purposes. The inert device is loaded with a non-irritant, non-toxic formulation payload. The device may also contain a non-toxic, water-soluble marker to allow immediate feedback on content delivery to target. The MK-4 device is marked "Inert." (Figure A-16)

Figure A-17. MK-9 Squad Inert OC Pepper Spray

25. MK-9 Squad Inert OC Pepper Spray: The MK-9 squad inert pepper spray has a range of up to 15 feet in a target-specific fog. It contains 8 to 10 half-second high volume bursts. It provides the service member with an inert canister for familiarization and training purposes. The inert device is loaded with a non-irritant, non-toxic formulation payload. The device may also contain a non-toxic, water-soluble marker to allow immediate feedback on content delivery to target. The inert MK-9 device is prominently marked "Inert" with a yellow sticker attached. (Figure A-17)

Figure A-18. MK-46 Platoon Inert OC Pepper Spray

26. MK-46 Platoon Inert OC Pepper Spray: The MK-46 platoon inert pepper spray has a range of up to 25 to 30 feet in a target-specific dispersed spray pattern. It contains up to 12 one-second high volume bursts. It provides the service member with an inert canister for familiarization and training purposes. The inert device is loaded win a non-irritant, non-toxic formulation payload. The device may also contain a non-toxic, water-soluble marker to allow immediate feedback on content delivery to target. The inert MK-46 device is prominently marked "Inert" with a yellow sticker attached. (Figure A-18)

Figure A-19. MK-46 Refill Inert OC Pepper Spray

27. MK-46 Refill Inert OC Pepper Spray: The MK-46 refill solution comes in a one-gallon container. (Figure A-19)

Figure A-20. FIST Training Suit

28. FIST Training Suit: The FIST training suit is made of closed-cell shock-absorbent foam that allows service members to hone their riot baton skills through realistic hand-to-hand engagement training. The FIST suit can absorb blows inflicted by the current expandable and wood riot baton. It provides protection for the head, face, hands, and legs, without significant degradation of the wearer's mobility. The FIST suit consists of a helmet, chest and back protector, bicep and forearm protectors, groin and buttock protectors, thigh and shin protectors, gloves, and carry bag. (Figure A-20)

NOTE: FIST is the current manufacturer of this training suit).

Figure A-21. Riot Training Strike Bag

29. Riot Training Strike Bag: The riot training strike bag gives the service member a strike bag for training and proficiency in open-hand control techniques. (Figure A-21)

Figure A-22. Caltrop

30. Caltrop: The caltrop is a field expedient area denial system. Securing approximately 50 caltrops to 550 military cord best employs it. This allows a service member to quickly toss/recover the system, using the caltrops to deny vehicle/pedestrian movement in a designated area. The four prongs are approximately two inches tall and will puncture tires or boots. (Figure A-22)

Figure A-23. Road Side Spike Strip

31. Road Side Spike Strip: The spike strip is designed to puncture a pneumatic tire as the vehicle makes contact with the spike strip. The spikes pivot into the tire and the tip-guards are pushed down. The spikes are imbedded deeply into the tire. The spikes are pulled out of the unit and remain in the tire. On the second revolution they will be pushed directly into the tire, allowing air to escape slowly without causing a blowout or sudden loss of control. The average tire will pick up 4 to 6 spikes with the deflation time of about 12 to 20 seconds. It allows a service member to quickly toss/recover the system, impeding vehicle movement in a designated area. It allows a force to provide additional physical security in areas where the caltrop cannot be tactically employed. (Figure A-23)

Figure A-24. High-Intensity Search Light

32. The high-intensity searchlight is a rugged, lightweight, hand-held, high-intensity spotlight for use during low-light conditions. Two different high-intensity lights were fielded. The Maxabeam light (shown on top) was fielded in 1997. It was replaced by the Nighthunter light (shown underneath) in 2000. (Figure A-24)

Figure A-25. Individual Light Holster (6P)

33. Individual Light Holster (6P): The 6P individual light holster is designed to hold the 6P individual light. (Figure A-25)

Figure A-26. 6P Individual Light

34. 6P Individual Light: The 6P individual light is issued to service members within the platoon. (Figure A-26)

Figure A-27. 6P Individual Light Accessory Kit

35. 6P Individual Light Accessory Kit: The 6P individual light accessory kit includes a spare light bulb installation kit. (Figure A-27)

Figure A-28. 3V Lithium Batteries

36. 3V Lithium Batteries: Two 3V Lithium Batteries are required to operate the 6P individual light. (Figure A-28)

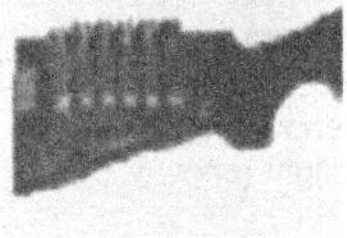

Figure A-29. 12-Gauge Buttstock Cuff

37. 12-Gauge Buttstock Cuff: The buttstock cuff is designed to fasten to the buttstock of most shotguns to hold seven 12-gauge cartridges for quick loading of the shotgun. (Figure A-29)

Figure A-30. 12-Gauge Utility Pouch

38. 12-Gauge Utility Pouch: The 12-gauge utility pouch is designed to hold twenty-five 12-gauge munitions and attaches to the service member's combat gear. (Figure A-30)

Figure A-31. 40-mm Carrying Pouch

39. 40-mm Carrying Pouch: The 40-mm carrying pouch is partitioned in a manner to provide the service member with a means of carrying not less than six 40-mm nonlethal rounds in separate compartments. It attaches to the service member's combat gear. (Figure A-31)

Figure A-32. Diversionary/Rubber Ball Grenade Pouch

40. Diversionary/Rubber Ball Grenade Pouch: The diversionary/rubber ball grenade pouch is an ammunition carrying pouch partitioned in a manner to provide the service member with a means of holding up to six rubber ball grenades or six flash bang grenades in separate compartments. (Figure A-32)

Figure A-33. 12-Gauge Mossberg 500A2/590 Shotgun

41. 12-Gauge Mossberg 500A2/590 Shotgun: The 12-gauge Mossberg 500A2 or 590 shotgun are the only 12-gauge shotguns currently authorized for firing 12-gauge nonlethal projectile munitions. The launching cup for the rubber ball grenade will attach to the Mossberg 500A2 Shotgun. The launch cup will not attach to the Mossberg 590 shotgun. Until the Joint Service Combat shotgun can effectively fire nonlethal 12-gauge rounds, the 12-gauge Mossberg 500A2 or 590 shotgun configuration meets the requirement of the NLW force. (Figure A-33)

Figure A-34. 12-Gauge M870 (modified) Version 2 (USCG)

42. 12-Gauge M870 (modified) Version 2 (USCG): The modified Remington 870 shotgun has a fixed synthetic stock and pistol grip. The pistol grip gives the shooter more control of the weapon and helps dissipate recoil from the shoulder. The synthetic materials used are more durable and lighter weight than the wood stocks used on the Standard M870. The barrel is fitted with a rifled choke tube for firing slugs (solid projectiles) to increase accuracy. It is also equipped with an Advanced Combat Optical Gunsight that was developed for use in "both-eyes-open" close quarters battle. The advantage is rapid target acquisition without the need to align front and rear sights. Batteries do not power the illumination sight; it contains tritium gas.

In the Coast Guard, a yellow "stock sock" will be used whenever the shotgun is loaded with nonlethal munitions. This removable nylon sleeve is used to identify the weapon as a nonlethal delivery system. (Figure A-34)

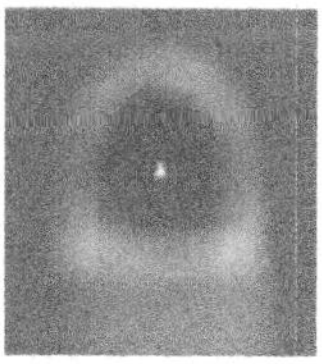

Figure A-35. Advanced Combat Optical Gunsight (ACOG) Reflex Sight (USCG)

43. Advancéd Combat Optical Gunsight (ACOG) Reflex Sight (USCG): Modified shotguns are equipped with this site. (Figure A-34)

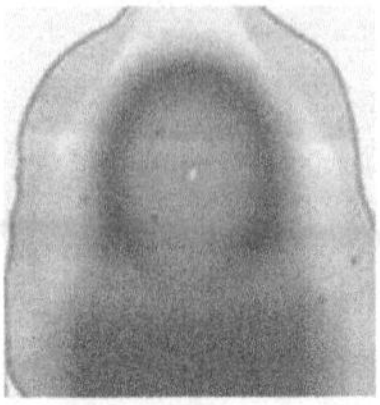

Figure A-36. ACOG Reflex Sight (USCG)

44. ACOG Reflex Sight (USCG): The dual-illuminated, Trijicon® Reflex sight was developed for the military for use in both-eyes-open close quarters battle (CQB). The advantage is quick target acquisition thus eliminating the need to align front and rear sights. Batteries do not power the illumination sight; it contains tritium gas. (Figure A-35)

Equipment Currently Being Fielded

1. 40-mm Crowd Dispersal Cartridge (CDC):

This direct fire, low-hazard, non-shrapnel-producing, multiple ball projectile provides friendly forces the capability to conduct crowd control. The CDC uses nonlethal, blunt trauma to stop, confuse, disorient, and/or temporarily incapacitate personnel and increase standoff distance between friendly forces and the hostile crowd. Primarily designed as an area effect munition, the CDC can be used on individuals. The M203 40-mm grenade launcher fires the CDC and contains 48, .48 caliber rubber balls. Effective range is 10 to 30 meters. (Figures A-37 and A-38)

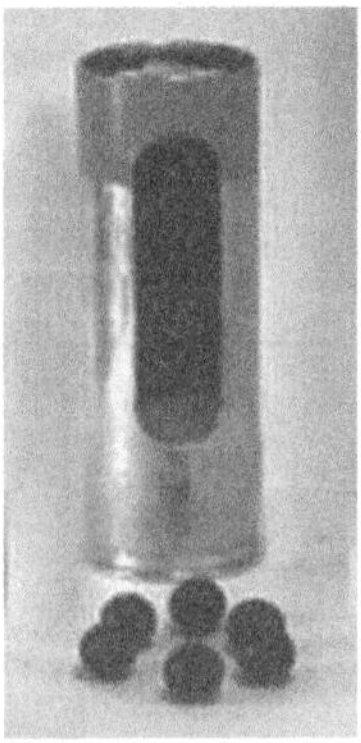

Figure A-37. Crowd Dispersal Cartridge

Figure A-38. Soldier Employing CDC with M203

2. Portable Vehicle Arresting Barrier (PVAB):

PVAB is a lightweight, portable, manually emplaced and recoverable barrier entrapment device to create, on command, a barrier that stops vehicles driven by known or suspected hostile forces with minimal damage to the vehicle or its occupants. It is used at dismount points, checkpoints, and roadblocks to prevent unauthorized access by wheeled vehicles into or out of areas (e.g. airfields, ports, command posts [CP]). Two individuals can easily emplace it, in approximately four hours. In stand-by mode, PVAB folds down into something similar to a speed bump (vehicles pass freely). In capture mode, the net springs up on two hydraulic pistons, to maximum height in less than a second. PVAB can be set up to span 12 to 24 feet and can stop a light truck (up to 7,500 pounds) at speeds up to 45 miles per hour, with a stopping distance of less than 200 feet PVAB can be containerized: 53.75 cubic feet (7 boxes, 4 large-3 small)—1,044 pounds on approximately one pallet. (Figures A-39 and A-40)

Figure A-39. PVAB in Action (Far View)

Figure A-40. PVAB in Action Near View

3. Modified Crowd Control Munition—Ground Emplaced (MCCM-GE):

The MCCM-GE (Figure A-41) is a nonlethal munition, housed in a casing similar to the current claymore mine, which will stop, confuse, disorient and/or temporarily incapacitate area targets/personnel at close range. The force can use it to provide crowd control by dispersing hostile crowds and force protection by integrating it into a layered perimeter defense. It delivers 600, .32 caliber rubber balls as well as a flash bang effect. The MCCM GE gives the commander or small unit leader greater flexibility to apply nonlethal force as a first line of defense against hostile noncombatants. Effective range is 5 to 15 meters. (Figure A-42)

Figure A-41. Modified Crowd Control Munition-Ground Emplaced (MCCM-GE)

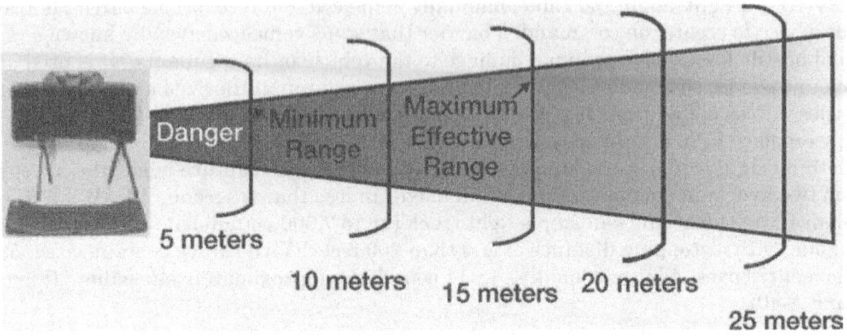

Figure A-42. MCCM-GE Range Chart

4. 66-mm Vehicle Mounted Nonlethal Grenade Launcher (VMNLGL):

The VMNLGL is a nonlethal grenade-launching device mounted on high mobility multipurpose wheeled vehicle (HMMWV). It is an indirect fire-support system that can deliver a 66-mm distraction device (flash-bang), CS, or blunt trauma (sting ball) payload. The distraction grenade uses a pyrotechnic charge for crowd control through audio and visual stimuli. The blunt trauma grenade uses a pyrotechnic charge to discharge 140, .32 caliber (cal.), rubber balls to achieve crowd control through audio, visual, and physical stimuli. The force uses VMNLGL to enhance the ability of friendly forces to conduct force protection without direct contact with hostile or threatening crowds. It enforces standoff distances and deters potential threats. It can disperse or subdue crowds, keep personnel from surrounding or climbing on patrol vehicles, slow or stop threatening personnel or vehicles by confusing, disorienting, and/or temporarily incapacitating individuals for capture. The grenade is fired from the grenade launchers traditionally found on tracked vehicles. Vehicle-launched grenades have three submunitions designed to ground burst. Effective range is 50 to 100 meters. An adjustable bracket allows engagement at ranges of 50 meters, 75 meters, and 100 meters. (Figure A-44)

Figure A-43. 66-mm Vehicle Mounted Nonlethal Grenade Launcher (VMNLGL)

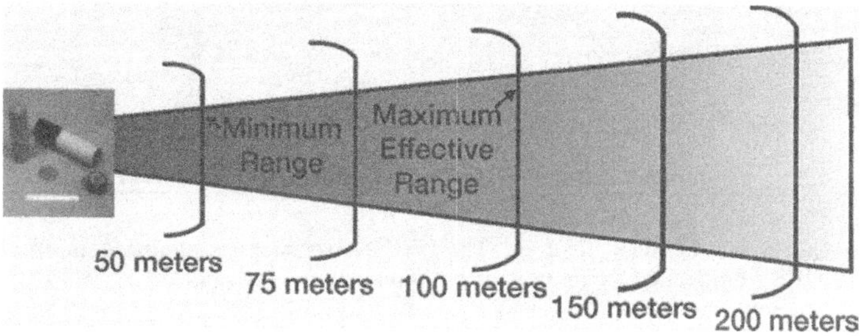

Figure A-44. VMNLGL Range Chart

5. Running Gear Entanglement System (RGES)

The Running Gear Entanglement System (RGES) is series of short, separate, looped polyethylene lines that are attached to a single, long polyethylene base line. This entanglement net fouls the propellers of planing hull small boats. The capability to manually deploy the net from a soft-sided bag has been fielded for USCG use in limited numbers. The USCG is in the process of incorporating three other variants of RGES into its inventory of nonlethal capabilities. These include aerial-delivered RGES, surface launched RGES to deploy from vessels, and static barrier RGES. The static barrier RGES has broad application into support homeland security, physical security and force protection missions. The surface launched RGES has direct application in the USCG's law enforcement mission wherever the requirement exists to stop a non-compliant vessel. Both of these systems are expected to be fielded in the fall/winter of CY02. The air launched version is currently still in the concept development phase and is not yet fully developed. (Figure A-45)

Figure A-45. Running Gear Entanglement System (RGES)

a. RGES Surface Delivery System: The USCG can use stern-mounted, manually deployed RGES from a cutter's small boat to stop slow-moving migrant-type craft. (Figure A-46)

(1) Uses COTS compressed air cylinder launcher.

(2) Removable, fix-mounted launcher on side of craft.

(3) Launch projects device ahead of target.

Figure A-46. RGES Surface Delivery System

b. RGES Barrier

(1) Provides static barrier around ship or other waterside asset.

(2) Provides a visible demarcation line, establishes zones, deters intrusions. (Figure A-47)

Figure A-47. RGES Barrier Employment

c. RGES Aerial Delivery System (ADS): RGES is aerial deployed from the ADS ahead of target aboard a H60 helicopter (Figure A-48). Device should float in the water ahead of target ensuring target runs over device. Device fouls propeller or rudder until target stops and awaits boarding.

(1) Uses COTS compressed air cylinder launcher

(2) Removable, fix-mounted launcher on helicopter

(3) Launch projects device ahead of target. (Figures A-49)

Figure A-48. RGES Aerial Delivery System (ADS)

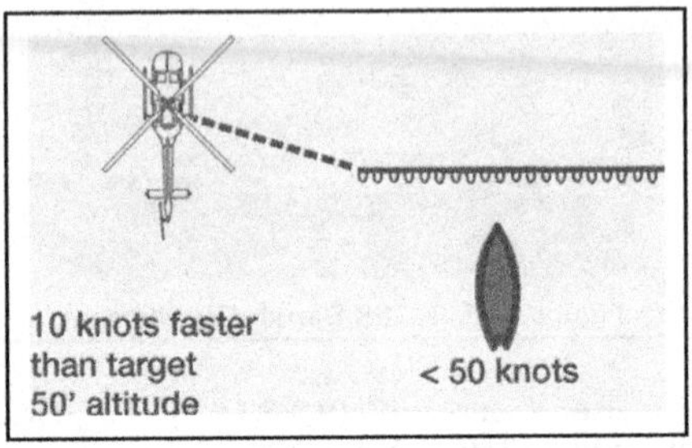

Figure A-49. RGES Aerial Delivery System (ADS)

6. Warp Cutter

a. Warp cutting is a nonlethal use of force, which has been safely and effectively employed by several countries to enforce fisheries laws in their respective exclusive economic zones (EEZ). The warp cutter is a device that closely resembles a grappling hook or small anchor. It has hardened steel blades attached to the crux of the grapnels. It is towed behind an enforcement vessel and across the warps of a non-compliant fishing vessel. As the cutting device engages the warps, it slices through them, separating the net from the fishing vessel.

b. The USCG uses the warp cutter to compel foreign vessels suspected of illegally fishing in the U.S. EEZ to comply with lawful orders to stop and be boarded. The USCG may cut the trawl warps of non-compliant vessels as a means of compelling them to stop for a boarding.

Figure A-50. Warp Cutter

Appendix B

Summary: Specialty Impact Munitions

12-Gauge Launch Cartridge

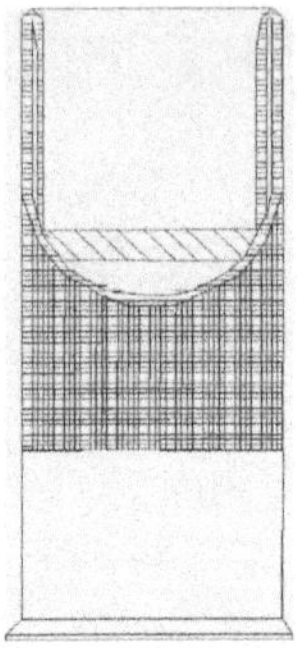

Using Services: USN, USMC
Status: Fielded 1998
DODIC: AA30
NSN: 1305-01-464-8389
Acquisition Activity: MARCORSYCOM—NSWC Crane, IN
Further Information:

Commercial: MARCORSYSCOM (703) 784-9395
NSWC Crane (812) 854-5801

Color....................................Clear case
Diameter..............................12-gauge
Projectile..............................N/A
Length..................................2.00 inches (51 mm)
Total Weight.........................13.6 grams
Propellant Weight.................46.3 grains (black powder)
Muzzle Velocity.....................100 feet/second (30.5 m/sec)—rubber ball grenade
Acceptance Accuracy..........200 feet (61 m) + 25 feet (7.6 m) at 30-degree launch angle using rubber ball grenade and 12-gauge launch cup

Projectile Specification: N/A

- Use 12-gauge launch round in combination with rubber ball grenade and 12-gauge launch cup to hurl a grenade a distance of 200 feet (61 m) + 25 feet (7.6 m) when launch angle is approximately 30 degrees.
- Chamber launch cartridge only after installing grenade in launch cup and pulling safety pin from grenade.

Figure B-1. 12-Gauge Launch Cartridge

12-Gauge Bean Bag Cartridge

Using Services: USN, USMC

DODIC: AA29

NSN: 1305-01-454-0191

Status: Fielded 1998

Acquisition Activity: MARCORSYCOM—NSWC
Crane, IN

Further Information:

Commercial: MARCORSYSCOM (703) 784-9395
NSWC Crane (812) 854-5801

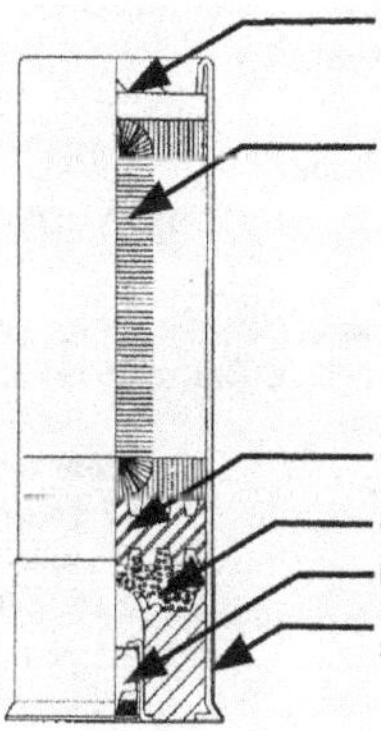

Color ....Clear or red case
Diameter ....12 gauge
Projectile ....1 bean bag
Length ....2.45 inches (62 mm)
Total Weight............................49.9 grams
Propellant Weight...................3.4 grains
Muzzle Velocity.......................300 + 25 feet/second (91.4 + 7.62 m/sec)
Acceptance Accuracy..........2 feet (.6 m) X 6 feet (1.8 m) at 100 feet (30.3 m) with Mossberg 500A2

Shotgun Projectile Specification:

Shape................................Square bag or tube sock
Material.............................Nylon or cotton-polyester
Filler..................................#9 lead shot
Filler weight......................44.4 grams (maximum)

- Fatal injuries are possible at employment distances of less than 20 feet (6 m)
- At distances of 20 feet (6 m) to 40 feet (12 m), target area should be center mass
- Training needs to reinforce that head shots are NOT ACCEPTABLE
- Do not skip fire this round! Round becomes unpredictable upon striking the ground
- Round is most effective against individually selected targets (point round)
- End wad has low light marker

Figure B-2. 12-Gauge Bean Bag Cartridge

12-Gauge Rubber Fin Stabilized Cartridge

Using Services: USN, USMC, USCG

Status: Fielded 1998

DODIC: AA31

NSN: 1305-01-454-0189

Acquisition Activity: MARCORSYCOM—NSWC Crane, IN

Further Information:

Commercial: MARCORSYSCOM (703) 784-9395
NSWC Crane (812) 854-5801

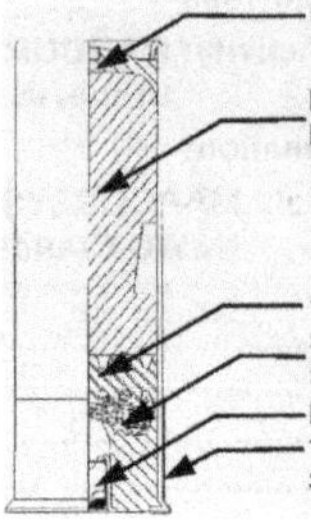

Color .... Clear or red case
Diameter .... 12 gauge
Projectile .... 1 rubber projectile
Length .... 2.45 inches (62 mm)
Total Weigh .... 9.1 grams
Propellant Weight................. 4.8 grains
Muzzle Velocity .... 530 + 65 feet/second (161.5 + 19.8 m/sec)
Acceptance Accuracy .... 2 feet (.6 m) X 6 feet (1.8 m) at 100 feet (30.3 m) with Mossberg 500 A2 shotgun
Projectile Specification: ...... Caliber 12 gauge
Material.................................. Rubber
Hardness .... Durometer 70-80 Shore "A" scale
Weight .... 7.8 grams
Temperature Limits (demonstrated):
Upper limit...125° F
Lower limit...0° F

Shipping Data: Hazard classification/division and storage compatibility group...1.4S
DOT Class .... ORM-D
DOT Markings Cartridge...... Small Arms
UNO serial number............... 0012

- Fatal injuries are possible at employment distances of less than 20 feet (6 m); USCG: 15 yards
- At distances of 20 feet (6 m) to 40 feet (12 m), target area should be center mass; USCG: 15 to 30 yards
- Training needs to reinforce that head shots are NOT ACCEPTABLE
- Do not skip fire this round! Round becomes unpredictable upon striking the ground
- Round is most effective against individually selected targets (point round)
- End wad has low light marking

Figure B-3. 12-Gauge Rubber Fin Stabilized Cartridge

40-mm Foam Rubber Baton Cartridge

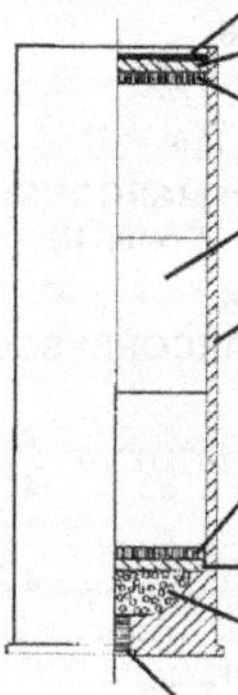

Using Services: USN, USMC

Status: Fielded 1998

DODIC: BA07

NSN: 1310-01-453-9168

Acquisition Activity: MARCORSYSCOM—NSWC Crane, IN

Further Information:

Commercial: MARCORSYSCOM (703) 784-9395
NSWC Crane (812) 854-5801

Color......................................**Aluminum (case)**
Diameter................................**40 millimeter**
Projectile...............................**3 foam batons**
Length...................................**4.78 inches (121.4 mm)**
Total Weight..........................**201 grams**
Propellant Weight.................**13.0 grains**
Muzzle Velocity.....................**325 + 35 feet/second (99 + 10.6 m/sec)**
Acceptance Accuracy..........**6 feet (1.8 m) X 3 feet (.9 m) at 50 feet (15.2 m) with M203**
Projectile Specifications:
Number............................**3**
Material............................**Foam rubber**
Hardness.........................**Durometer 25 + 10 on Shore "A" scale**
Weight..............................**250 to 280 grains**

- **Fatal injuries are possible at employment distances of less than 20 feet (6 m)**
- **At distances of 20 feet (6 m) to 40 feet (12 m), target area should be center mass**
- **Training needs to reinforce that head shots are NOT ACCEPTABLE**
- **Round is most effective against point targets (individual personnel)**
- **End wad has low light marker**

Figure B-4. 40-mm Foam Rubber Baton Cartridge

40-mm Rubber Ball Cartridge

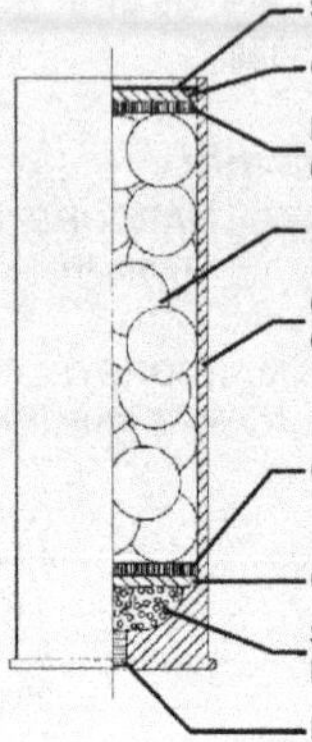

Using Services: USN, USMC

Status: Fielded 1998

DODIC: BA08

NSN: 1310-01-453-9154

Acquisition Activity: MARCORSYSCOM—NSWC Crane, IN

Further Information:

Commercial: MARCORSYSCOM (703) 784-9395

Color	Aluminum (case)
Diameter	40 millimeter
Projectile	24 rubber balls
Length	4.78 inches (121.4 mm)
Total Weight	201 grams
Propellant Weight	7.3 grains
Muzzle Velocity	325 + 35 feet/second (99 + 10.6 m/sec)
Acceptance Accuracy	6 feet (1.8 m) x 9 feet (2.7 m) at 50 feet (15.2 m) with M203NSWC Crane (812) 854-5801

Projectile Specifications:

Caliber	60
Number	22 (minimum)
Material	Rubber
Hardness	Durometer 60 + 10 on Shore "A" scale
Weight	2.2 grams each

- Fatal injuries are possible at employment distances of less than 20 feet (6 m)
- At distances of 20 feet (6 m) to 40 feet (12 m), target area should be center mass
- Training needs to reinforce that head shots are NOT ACCEPTABLE
- Round is most effective against area targets (multiple personnel)
- End wad has low light marker

Figure B-5. 40-mm Rubber Ball Cartridge

Rubber Ball Grenade

Using Services: USN, USMC

Status: Fielded 1998

DODIC: GG04

NSN: 1310-01-454-0132

Acquisition Activity: MARCORSYSCOM—NSWC Crane, IN

Further Information:

Commercial: MARCORSYSCOM (703) 784-9395
NSWC Crane (812) 854-5801

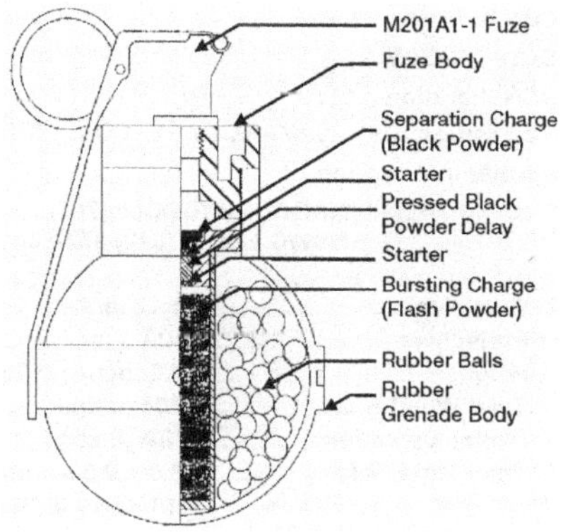

Color.......................................Black with white lettering
Diameter.................................3.12 inches (79 mm)
Projectiles.............................100 rubber balls (minimum)
Height.....................................5.25 inches (133.4 mm)
Total Weight..........................290 grams
Explosive Weight...................59 grams
Total Time Delay...................3.0 + 0.5 seconds

Employment Distance:

Hand Thrown.......................50 feet (15.2 m)
Shotgun Launched............200 feet (61 m)

Projectile Specifications:

Number...................................100 (minimum)
Material..................................Rubber
Hardness.............................Durometer 70 to 80 on Shore "A" scale
Diameter...............................0.25 inches (6.4 mm) (minimum)

- Fuze separates from the body 1.3 to 1.7 seconds after primer ignition.
- Second delay causes burster charge detonation 3.0 + 0.5 seconds average after primer ignition. Rubber balls are then dispersed in a 50-feet (15-m) circular pattern.
- 12-gauge shotgun with launch cup and launch cartridge can launch grenade 200 feet (61 m) + 25 feet (7.6 m)

Figure B-6. Rubber Ball Grenade

Practice Grenade Body

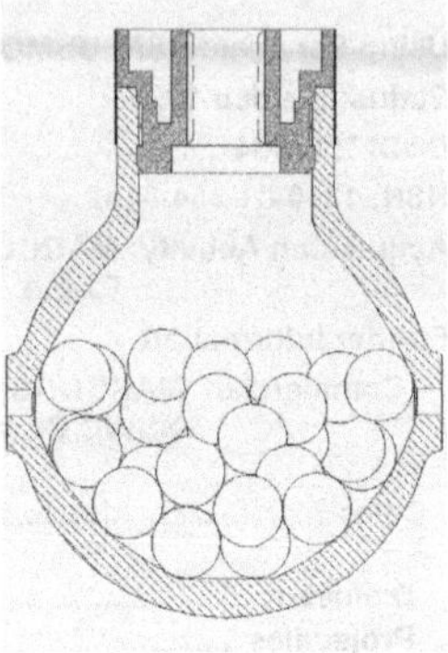

Using Services: USMC

Status: Fielded 2002

DODIC: GG05

NSN: Pending

Acquisition Activity: MARCORSYSCOM – NSWC Crane, IN

Further Information:

Commercial: MARCORSYSCOM (703) 784-9395
NSWC Crane (812) 854-5801

Diameter................................3.12 inches (79 mm)
Projectiles ....N/A
Height ....4.2 inches (106 mm)
Total Weight..........................208 grams
Explosive Weight..................N/A (inert)
Total Time Delay...................3.0 + 0.5 seconds when used with practice grenade fuze

Employment Distance:
Hand Thrown...................50 feet (15.2 m)
Shotgun Launched ....200 feet (61 m) using Mossberg launch cup and 12-gauge launch cartridge (AA30).

- Practice grenade body is reusable up to 25 times and is a component of a practice grenade assembly, consisting of the body and a practice grenade fuze.
- Practice grenade assembly accurately replicates the GG04 rubber ball grenade and is used for training purposes.

Figure B-7. Practice Grenade Body

Practice Grenade Fuze

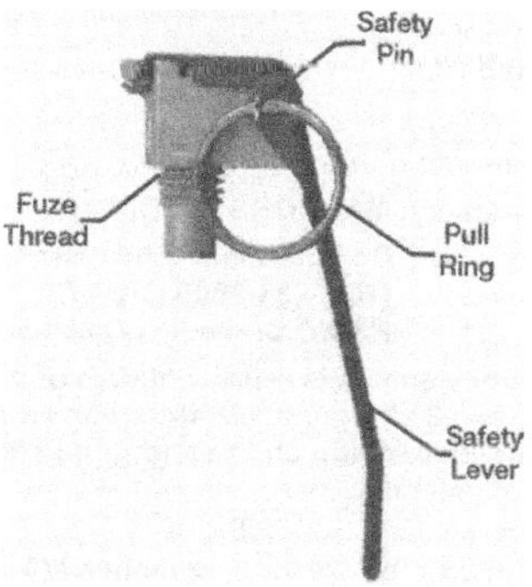

Using Services: USMC

Status: Fielded 2002

DODIC: Pending

NSN: Pending

Acquisition Activity: MARCORSYSCOM—NSWC Crane, IN

Further Information:

Commercial: MARCORSYSCOM (703) 784-9395
NSWC Crane (812) 854-5801

Color	Blue safety lever with brown stripe
Projectiles	N/A
Height	4.25 inches (107 mm)
Total Weight	82 grams
Explosive Weight	Not available
Total Time Delay	3.0 + 0.5 seconds

Employment Distance when used with GG05 Practice Grenade Body:

Hand Thrown	50 feet (15.2 m)
Shotgun Launched	200 feet (61 m)—using Mossberg launch cup and 12-gauge launch cartridge (AA30).

- Employ practice grenade fuze with the practice grenade body to create a practice grenade assembly consisting of the body and a practice grenade fuze.
- Practice grenade assembly accurately replicates the GG04 rubber ball grenade and is used for training purposes.

Figure B-8. Practice Grenade Fuze

Nonlethal Grenade Launch Cup

Using Service: USMC

Status: Fielded 2002

DODIC: N/A

NSN: Pending

Acquisition Activity: MARCORSYSCOM—NSWC Crane, IN

Further Information: Commercial: MARCORSYSCOM
(703) 784-2006 Ext. 2733
NSWC Crane (812) 854-5801

NOTE: Cup body shown is representative of cup bodies used on all shotguns. Adapter shown is subject to change depending on the shotgun used and design improvements.

Color .... Black
Diameter (internal) .... 3.12 inches (79 mm)
Length (without adapter) ..... 4 inches (102 mm)
Weight .... 397 grams (maximum)

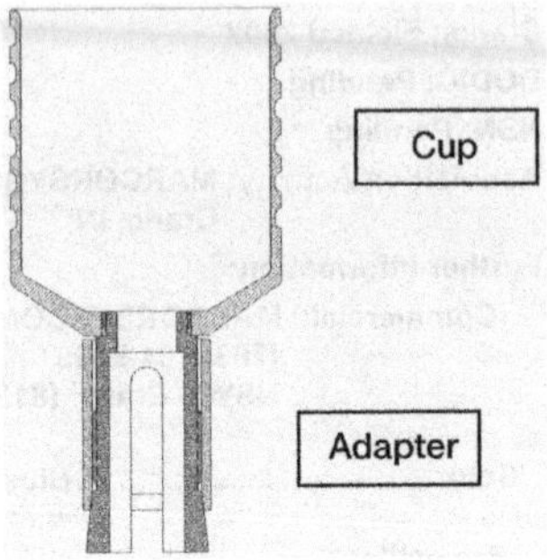

Cross Section View

- Use launch cup in concert with AA30 launch cartridge and either the GG04 rubber ball grenade or GG05 practice grenade assembly
- Attach launch cup and remove from the shotgun without using tools
- Launch cup can withstand accidental detonation of rubber ball grenade without fragmenting

Figure B-9. Nonlethal Grenade Launch Cup

Diversionary Charge, MK 141, MOD 0

Using Services: USN, USMC

Status: Fielded 1997

DODIC: DWBS

NSN: 1375-01-387-5748

Acquisition Activity: MARCORSYSCOM—NSWC Crane, IN

Further Information:

Commercial: MARCORSYSCOM (703) 784-9395
NSWC Crane (812) 854-3934

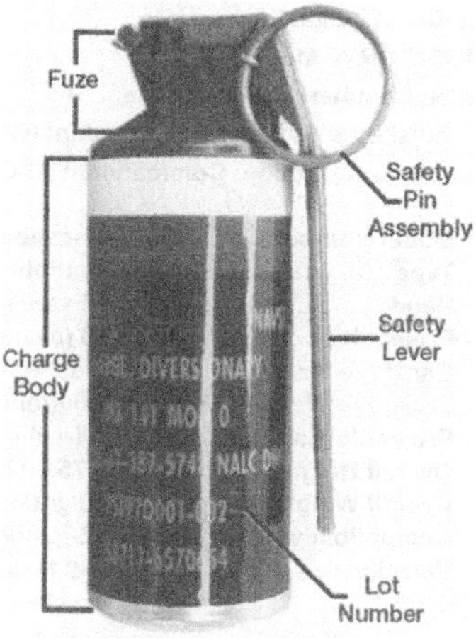

Color	Brown with white stripe
Diameter	1.77 inches (45 mm)
Projectile	N/A
Length	5.214 inches (132.4 mm)
Total Weight	99.8 grams
Net Explosive Weight	17.8 grains
Light Output	2 million candlepower for 4 milliseconds
Noise Output	176 to 185 decibels
Material	Molded and machined polyurethane
Time Delay:	
Initial	1.26 seconds (minimum)
Total	1.34 to 1.84 seconds

- Diversionary charge can cause serious injury if held at time of detonation
- Light and noise most effective in confined space to stun individuals
- Most often employed for hostage rescue

Figure B-10. Diversionary Charge, MK 141, MOD 0

Cartridge, 12-Gauge Rubber Sting Ball™

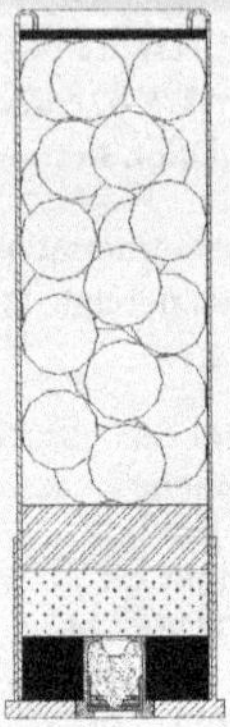

Using Services: USCG
Status: Fielded
DODIC: N/A **Model No.** 2552
NSN: Commercial Procurement
Acquisitions Activity: Commandant (G-OCU)
Further Information: Commandant (G-OCU) (202) 267-1522

Model Number	31-caliber sting-ball
Type	Multiple projectile rubber pellets
Range	10 yards (9 m)
Projectiles	18 (average)
Cartridge Material	Plastic
Projectile Weight	.35 grams each
Projectile Material	31-caliber rubber pellets
Overall Height	2.75 inches (70 mm)
Overall Weight	18 grams
Compatibility	12-gauge shotguns
Velocity	400 feet/second

USCG

- Fatal injuries are possible at employment distances of less than 5 yards.
- At distances of 20 feet (6 m) to 40 feet (12 m), target area should be center mass
- This round is most effective against area targets with a maximum effective range of 15 yards
- Training needs to reinforce that head shots are NOT ACCEPTABLE
- Round is most effective against area targets (multiple personnel)
- Stock Sock: a yellow "stock sock" will be used whenever the modified 12-gauge Remington 870 shotgun is loaded with nonlethal munitions. This removable nylon sleeve is used to identify the weapon as a nonlethal delivery system.

- All specifications are average and are subject to change without notice
- 2552SPEC

Figure B-11. Cartridge, 12-Gauge Rubber Sting Ball™

Cartridge, 12-Gauge Sting Ball™ High Velocity

Using Services: USCG
Status: Fielded
DODIC: N/A Model No. 2553
NSN: Commercial Procurement
Acquisitions Activity: Commandant (G-OCU)
Further Information: Commandant (G-OCU) (202) 267-1522

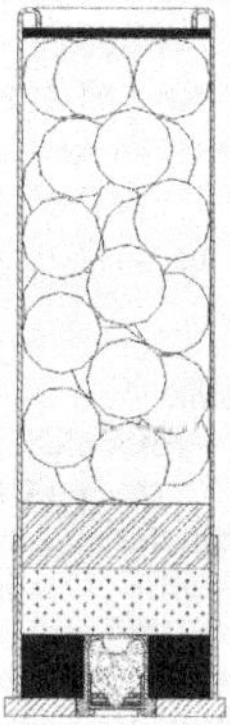

Model Number	31-caliber sting-ball
Type	Multiple projectile rubber pellets
Range	10 yards (9 m)
Projectiles	18 (average)
Cartridge Material	Plastic
Projectile Weight	.35 grams each
Projectile Material	.31-caliber rubber pellets
Overall Height	2.4 to 2.5 inches (61 – 63 mm)
Overall Weight	18 grams
Compatibility	12-gauge shotguns
Velocity	900 feet/second

USCG

- Fatal injuries are possible at employment distances of less than 5 yards.
- At distances of 20 feet (6 m) to 40 feet (12 m), target area should be center mass
- Maximum effective range is 15 yards
- Training needs to reinforce that head shots are NOT ACCEPTABLE
- Round is most effective against area targets (multiple personnel)
- Stock Sock: a yellow "stock sock" will be used whenever the modified 12-gauge Remington 870 shotgun is loaded with nonlethal munitions. This removable nylon sleeve is used to identify the weapon as a nonlethal delivery system.

- All specifications are average and are subject to change without notice.
- 2553SPEC

Figure B-12. Cartridge, 12-Gauge Sting Ball™ High Velocity

M1006, 40-mm Nonlethal Cartridge (Sponge Grenade)

Using Services: U.S. Army
Status: Materiel Released April 2000
DODIC: BA06
NSN: 1310-01-452-1190
Acquisition Activity: Acquisition Center, Picatinny Arsenal, NJ
Further Information:
Commercial: Picatinny Arsenal NL Systems Integrator: (973) 724-6948
Product Director: (973) 724-6283

Color:
Projectile .... Black, olive green (foam rubber)
Cartridge Case .... Opaque
Diameter .... 40 millimeter
Length .... 5.272 inches (13.4 cm) (maximum)
Total Weight .... 68 grams (0.15 lbs)
Propellant Weight .... 2 grains
Muzzle Velocity .... 265 + 20 feet/second (81 + 6 m/sec)
Acceptance Accuracy .... >85 % accuracy at 30 meters (E-silhouette) and >60 % accuracy at 50 meters (E-silhouette) when fired from an M16A2/M203 or a M4/M203
Engagement Range .... 10 meters to 50 meters
Projectile Specification:
Shape .... 40-mm bullet shaped foam rubber
Material .... Foam rubber
Weight .... 28.5 grams (.06 lbs)

- Fatal injuries are possible at employment distances of less than 32.8 feet (10 m)
- At distances of 10 meters to 50 meters, target area should be center mass
- Training needs to reinforce that head shots are NOT ACCEPTABLE
- Do not skip fire this round! Round becomes unpredictable upon striking the ground
- Round is most effective against individually selected targets (point round)

Figure B-13. M1006, 40-mm Nonlethal Cartridge (Sponge Grenade)

M1012, 12-Gauge Shotgun: Nonlethal, Point Control Cartridge

Using Services: U.S. Army

Status: Type Classified, June 2000

DODAC: AA51

NSN: 1305-01-470-2405

Acquisition Activity: Acquisition Center, Picatinny Arsenal, NJ

Further Information:

Commercial: Picatinny Arsenal NL Systems Integrator: (973) 724-6948
Product Director: (973) 724-6086

Color:

Cartridge Case: .... Clear

Diameter .... 12 gauge

Length .... 2.45 inches (62.2 mm)

Total Weight .... 500 grains

Propellant Weight .... 6.6 grains smokeless (Red Dot)

Muzzle Velocity .... 500 feet/second (152.4 m/sec)

Acceptance Accuracy .... 80 % accuracy at 10 and 20 meters (E-silhouette) fired out of Mossberg 500, Mossberg 590, and Winchester 1200

Engagement range .... 10 to 20 meters

Projectile Specification:

Shape .... Fin stabilized 'bomblet' shape

Material .... Rubber

Weight .... 0.2 ounces (5.8 gm)

Hardness .... 75 Durometer "A" Scale

- Fatal injuries are possible at employment distances of less than 33 feet (10 m)
- At distances of 10 to 20 meters, target area should be center mass
- Training needs to reinforce that head shots are NOT ACCEPTABLE
- Round is most effective against individually selected targets (point round)

Figure B-14. M1012, 12-Gauge Shotgun: Nonlethal, Point Control Cartridge

M1013, 12-Gauge Shotgun: Nonlethal, Crowd Dispersal Cartridge

Using Services: U.S. Army
Status: Type Classified June 2000
DODAC: AA52
NSN: 1305-01-470-2139
Acquisition Activity: Acquisition Center, Picatinny Arsenal, NJ
Further Information:
Commercial: Picatinny Arsenal NL Systems Integrator: (973) 724-6948
Product Director :(973) 724-6086

Color:
- Projectile .... Black rubber
- Cartridge Case .... Clear

Diameter .... 12 gauge
Projectile .... 18 rubber balls
Length .... 2.37 inches (60.2 mm)
Total Weight .... 500 grains (32.4 gm)
Propellant Weight .... 14 grains smokeless (Red Dot)
Muzzle Velocity .... 900 feet/second (274 m/sec)
Acceptance Accuracy .... 80 % accuracy for a 2-by-3-meter target at 10 and 20 meters fired out of Mossberg 500, Mossberg 590, and Winchester 1200
Engagement range .... 10 to 20 meters
Projectile Specification:
- Shape .... 32-gauge (0.32 in) ball
- Material .... PVC rubber compound
- Weight .... 0.01 ounces (0.25 gm)
- Hardness .... 75 Durometer "A" Scale

- Fatal injuries are possible at employment distances of less than 32.8 feet (10 m)
- At distances of 10 to 20 meters, target area should be center mass
- Training needs to reinforce that head shots are NOT ACCEPTABLE

Figure B-15. M1013, 12-Gauge Shotgun: Nonlethal, Crowd Dispersal Cartridge

M1029, 40-mm Nonlethal Crowd Dispersal Grenade

Using Services: U.S. Army

Status: Type Classified May 2001

DODIC: BA13

NSN: 1310-01-475-0628

Acquisition Activity: Acquisition Center, Picatinny Arsenal, NJ

Further Information:

Commercial: Picatinny Arsenal NL Systems Integrator: (973) 724-6948
Product Director: (973) 724-6283

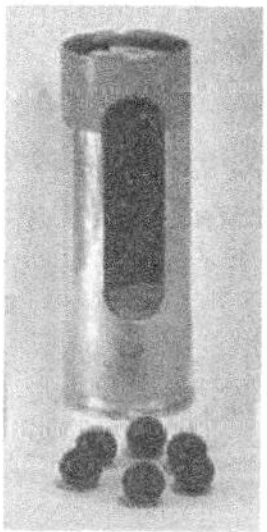

Color:
- Projectile .... Black rubber
- Cartridge Case .... Aluminum

Diameter .... 40 millimeter
Projectile .... 48 rubber balls
Length .... 4.8 inches (12.2 cm)
Total Weight.......................... 202.22 grams (7.0 oz)
Propellant Weight................. 8.5 grains smokeless
Muzzle Velocity..................... 450 feet/second
Acceptance Accuracy.......... > 90 % hit accuracy 4 out of 5 (shoulder to shoulder) E-silhouettes target at 30 meters and when fired from an M16A2/M203 or a M4/M203

Projectile Specification:
- Shape .... 48-gauge ball (.48 in diameter)
- Material............................. Rubber
- Weight .... 1.3 grams (.046 oz)
- Hardness........................... Durometer 60, scale "A"

- Fatal injuries are possible at employment distances of less than 32.8 feet (10 m)
- At distances of 10 to 30 meters, target area should be center mass of group of individuals
- Round is most effective when used to disperse groups of individuals

Figure B-16. M1029, 40-mm Nonlethal Crowd Dispersal Grenade

L96A1*, Grenade, Discharger: Antiriot, Irritant, CS

Using Services: U.S. Army, USMC

Status: Materiel Release, 4 January 2002

DODIC: FZ14

NSN: 1330-01-459-4018

Acquisition Activity: Acquisition Center, Picatinny Arsenal, NJ

Further Information:

Commercial: Picatinny Arsenal NL Systems Integrator: (973) 724-6948
Project Officer: (410) 436-2139

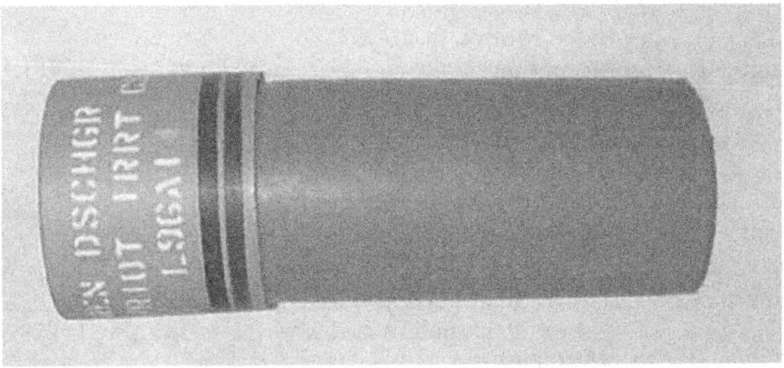

Color .... Light gray with red and brown bands
Diameter .... 66 millimeters
Projectile .... 23 canisters filled with CS compound
Length .... 185 millimeters (7.28 in)
Total Weight .... 568 grams (1.25 lbs)
Propellant Weight .... 3.8 grams (.008 lbs)
Muzzle Velocity .... 35.8 meters/second (117.5 ft/s)
Acceptance Accuracy .... 65 to 95 meters
Deployment Range .... 65 to 95 meters
Projectile Specification:

NOTE: It is not accurate to refer to the submunitions of the L96A1 grenade as projectiles. The antiriot material (CS) is released from the submunitions as they burn on the ground.

Shape .... Cylindrical
Material .... Aluminum
Weight .... 11.3 grams (.025 lbs)

* The L97A1 Grenade is the training grenade for the L96A1 Grenade.

Figure B-17. L96A1, Grenade, Discharger: Antiriot, Irritant, CS

L97A1, Grenade, Discharger: Antiriot, Practice

Using Services: U.S. Army, USMC
Status: Material Released, 4 January 02
DODIC: FZ15
NSN: 1330-01-459-4032
Acquisition Activity: Acquisition Center,
Picatinny Arsenal, NJ
Further Information:
Commercial: Picatinny Arsenal NL Systems
Integrator: (973) 724-6948
Project Officer:
(410) 436-2139)

Color ....Blue with brown and green bands
Diameter ....66 millimeters
Projectile ....23 canisters filled with Cinnamic Acid (CA)
Length ....185 millimeters (7.28 in)
Total Weight..........................568 grams (1.25 lbs)
Propellant Weight.................3.8 grams (.008 lbs)
Muzzle Velocity.....................35.8 meters/second (117.5 ft/s)
Acceptance Accuracy..........65 to 95 meters
Deployment Range...............65 to 95 meters
Projectile Specification:

NOTE: It is not accurate to refer to the submunitions of the L97A1 grenade as projectiles. The antiriot, practice material (CA) is released from the submunitions as they burn on the ground.

Shape..............................Cylindrical
Material............................Aluminum
Weight.............................11.3 gram (.025 lbs)

- CA smoke can cause irritation to the eyes, mucous membranes, and skin. Personnel entering a CA cloud must wear appropriate face protection such as an M17- or M40-series protective mask.
- The burning grenade canisters could start a fire in dry grasses. Keep fire extinguisher available.
- During firing training, personnel within 16 meters must wear single hearing protection. This includes personnel within vehicle.
- Do not fire grenades when personnel or equipment are within 150 meters covering a 90-degree arc around a firing discharger.
- Do not fire grenades at the 50- or 75-meter position on the M315 adjustable aiming bracket. Fire only at the 100-meter setting.

Figure B-18. L97A1, Grenade, Discharger: Antiriot, Practice

Grenade, Launcher: Nonlethal, Distraction, M98

Using Services: U.S. Army, USMC
Status: Type Classified, 17 October 2001
DODIC: FZ16
NSN: 1330-01-484-7773
Acquisition Activity: Acquisition Center, Picatinny Arsenal, NJ
Further Information: Commercial: Picatinny Arsenal NL Systems Integrator: (973) 724-6948
Project Officer: (410) 436-2139

Color ....Green and black with brown band
Diameter ....66 millimeters
Projectile ....3 bursting submunitions, with pyrotechnic charges for audio and visual stimuli
Length ....25.25 centimeters (9.94 in)
Total Weight..........................725.7 grams (1.6 lbs)
Propellant Weight.................5.0 grams (.011 lbs)
Muzzle Velocity.....................48.9 meters/seconds (160.4 ft/s)
Acceptance Accuracy..........100 to 120 meters
Deployment Radius.............10 to 15 meters
Projectile Specification:

NOTE: It is not accurate to refer to the submunitons of the M98 grenade as projectiles. These submunitions are ground bursting and the payload of inert ballast is just for flight stabilization and to deliver the audio/visual stimuli to the required range.

Filler ....Inert Ballast

- There is a remote chance of starting a fire. Keep fire extinguisher available.
- Personnel within 17 meters of a firing discharger must wear single hearing protection. This includes personnel within vehicle.
- Avoid looking directly at a bursting grenade; it could result in temporary loss of vision, lasting 2 to 5 minutes.
- Personnel within 35 meters of an operating launcher or 0.5 meters of a bursting canister should wear safety or ballistic-type eye protection, military clothing with long sleeves and elevated shirt collar, and helmet headgear.
- Do not handle or load M98s within 50 meters of a frequency transmitter emitting at 175 MHz, 448 MHz, 503 MHz and 1430 MHz.
- Do Not fire grenades when personnel or equipment are within 160 meters covering a 90-degree arc around a firing discharger.

Figure B-19. Grenade, Launcher: Nonlethal, Distraction, M98

M99, Grenade, Launcher: Nonlethal, Blunt Trauma

Using Services: U.S. Army, USMC

Status: Type Classified—17 October 2001

DODIC: FZ17

NSN: 1330-01-484-7775

Acquisition Activity: Acquisition Center, Picatinny Arsenal, NJ

Further Information:

Commercial: Picatinny Arsenal NL Systems Integrator: (973) 724-6948
Project Officer: (410-436-2139)

Color.....................................Green and black with brown band
Diameter................................66 millimeters
Projectile...............................3 bursting submunitions, each with 140 balls
Length...................................25.25 centimeters (9.94 in)
Total Weight..........................725.7 grams (1.6 lbs)
Propellant Weight.................5.0 grams (.011 lbs)
Muzzle Velocity.....................48.9 meters/second (160.4 ft/s)
Acceptance Accuracy..........100 to 120 meters
Deployment Radius..............5 to 10 meters
Projectile Specification:

NOTE: It is not accurate to refer to the submunitions of the M99 grenade as projectiles. These submunitions are ground bursting and the payload of inert ballast is just for flight stabilization and to deliver the audio/visual stimuli to the required range.

Shape...............................32-caliber (5/16 in) ball
Material............................PVC
Weight..............................0.40 grams (.014 oz)
Hardness.........................90 Durometer

- There is a remote chance of starting a fire. Keep fire extinguisher available.
- Personnel within 17 meters of a firing discharger must wear single hearing protection. This includes personnel within vehicle.
- Avoid looking directly at a bursting grenade; it could result in temporary loss of vision, lasting 2 to 5 minutes
- Personnel within 35 meters of an operating launcher or 0.5 meters of a bursting canister should wear safety or ballistic-type eye protection, military clothing with long sleeves and elevated shirt collar, and helmet headgear.
- Do not handle or load M98s within 50 meters of a frequency transmitter emitting at 175 MHz, 448 MHz, 503 MHz and 1430 MHz.
- DO NOT fire grenades when personnel or equipment are within 160 meters covering a 90-degree arc around a firing discharger.
- To minimize risk of eye injury, use the M99 primarily against crowds of adults.

Figure B-20. M99, Grenade, Launcher: Nonlethal, Blunt Trauma

Appendix C

Interservice Nonlethal Individual Weapons Instructor Course

1. Introduction

This appendix provides an overview of the Interservice Nonlethal Individual Weapons Instructor Course (INIWIC). The INIWIC is a "train-the-trainer" course taught in 15 subcourses and divided into four sections. Section I provides INIWIC descriptive data that includes the purpose, scope, peacetime and mobilization training requirements, student prerequisites, and equipment/ammunition requirements necessary to successfully complete the course. Section II lists the total training days and hours required per academic subcourse. Section III gives a detailed description of each subcourse and terminal learning objectives that each student must meet to complete the course. Section IV is a listing of 15 annexes that comprise the INIWIC Program of Instruction (POI). Access the entire INIWIC, with history, updates, and lesson plans at the following Internet address:
http://www.wood.army.mil/mcdet/niwic/niwic.htm.

2. Course Descriptive Data

a. Course Title: Interservice Nonlethal Individual Weapons Instructor Course

b. Location: Marine Corps Detachment, Fort Leonard Wood, Missouri

c. Marine Corps service school code: A16H5A3

d. Other service course number: USAF—L5AZA3P071019

e. Military assistance program articles and service list number: P166810

f. Purpose: To certify selected DOD/DOT service persons as nonlethal individual weapons instructors.

g. Scope: This is a train-the-trainer course designed for E-4s and above. The objective of this course is to prepare students to train individuals and groups in the utilization and employment of NLW and to serve as the unit NLW subject matter expert. This course is designed for all occupational specialties.

h. Length: 10 training days

i. Curriculum Breakdown: 100 total hours.

(1) 2.00 Demonstration.

(2) 24.00 Demonstration/practical exercise.

(3) 22.50 Lecture.

(4) 5.00 Lecture/demonstration.

(5) 4.00 Live-fire exercise.

(6) 26.50 Practical application.

(7) 6.00 Performance exam.

(8) 4.00 Written exam.

(9) 6.00 Administrative time.

j. Maximum Class Capacity: 30.

k. Optimum Class Capacity: 25.

l. Minimum Class Capacity: 15.

m. Class Frequency: 12.

n. Prerequisites: Students must be selected by commanders to be nonlethal weapons instructors and be an E-4 or above.

o. MOS Received: None.

p. Quota Control: Formal Schools Branch, Training and Education Command, Quantico, Virginia.

q. Funding: Service or command funded.

r. Reporting Instructions: All students report with orders to the Fort Leonard Wood, Missouri, Billeting Office, Bldg # 470 no later than 2400 on the day prior to class convening. Government quarters are available; however, messing is not.

s. Instructor Staffing: Two USMC instructors and two U.S. Army instructors.

t. School Overhead: Course OIC and chief instructor

Training Support: Training support is provided by the USMC for all common training support items. Service-specific training items are provided by the respective service.

Table C-1. Required Equipment For INIWIC

Item	*Quantity*
Personnel Protectors:	
Non-Ballistic Riot Face Shield	40
Non-Ballistic Full Length Riot Shield	40
Non-Ballistic Shin Guards	40
Ballistic Riot Face Shield	20
Ballistic Full Length Riot Shield	20
Ballistic Shin Guards	20
Kevlar Helmets	40
Flak Jacket/Body Armor	40
Personnel Effectors:	
Disposable Handcuffs (Flex Cuffs)	300
12-Gauge Area (rubber ball)	1400
12-Gauge Beanbag	1400
12-Gauge Fin Stabilized	1400
12-Gauge Flash-bang	1400
12-Gauge OO Buck	xxxx
12-Gauge 7 1/2 shot	xxxx
12-Gauge Launching Cup Cartridge	1400
40-mm Area (rubber ball)	1400
40-mm Foam Baton (3 foam sub-munitions)	1400
40-mm Sponge Grenade	1400
MK141 Diversionary Grenade	600

Table C-1. Required Equipment For INIWIC

Item	***Quantity***
M84 Reloadable Diversionary Grenade	600
Sting Ball Hand Thrown or Shot launched Grenade	600
Mossberg 500A2 Shotgun Launch Cup	15
M203 Grenade Launcher	15
Mossberg 500 3" Shotgun	15
36" Riot Oak Baton w/Belt Ring	40
Monadnock® 24-36" Expandable Baton w/holster	40
M36 RCA Dispenser	40
Mk-4 Personal OC Dispenser	40
Mk-9 Team OC Dispenser	40
Mk-46 High Capacity High Output OC Dispenser	20
MK-46 Hardware Kit (Refill Kit)	10
M33A1 Squad Riot Control Agent Dispenser	10
Modular Crowd Control Munitions Ground Emplaced (MCCM-GE)	20
66-mm Vehicle Launched Nonlethal Grenade (VLNLG)	3
66-mm NL munitions	30
Mission Enhancers:	
Portable Bullhorn	5
40-mm Carrying Pouch	40
Diversionary Rubber Ball Grenade Pouch	40
12-Gauge Butt Cuff carriers	40
12-Gauge Utility 25-Round Pouch	40
Mk-4 Individual Carrying Pouch	40
Portable Public Address System w/Microphone	3
12-Volt Hand-Held Spotlight	5
Night Hunter High Intensity Light System w/Kit	5
12-Gauge High Intensity Light Set	5
Caltrops	5
Road Side Spike Strip	5
Portable Vehicle Arrest Barrier (PVAB)	3
Training Devices:	
Impact Training Suit	6
Striking Bags	15
Training Batons (24 in)	40
Training Batons (36 in)	40
Practice Grenade Body	40
Practice Grenade Fuse	540
• OPFOR Training Munitions (foreign objects)	xxxx
• Target Requirements (full or partial torso)	40
MK-4 Personal OC Dispenser (training)	300
MK-9 Team OC Dispenser (training)	300
Training Disposable Handcuffs (flex cuffs)	100
12-Gauge Dummy Rounds	100

Table C-2. INIWIC Class V Requirements

Type	*#Per Student*	*# Demo Rounds*	*Total Per Class*	*Total Per Year*
12-Gauge Area (rubber ball)	5	5	155	1400
12-Gauge Beanbag	5	5	155	1400
12-Gauge Fin Stabilized	5	5	155	1400
40-mm Area (rubber ball)	5	5	155	1400
40-mm Foam Baton	5	5	155	1400
Sting Ball Hand Thrown or Shotgun Launched Grenades	2	2	62	600
MK141 Diversion Devices	2	2	62	600

3. Training Days Summary

This course consists of 10 training days. The curriculum is broken down into three categories.

Table C-3. INIWIC Curriculum

NLW Tactics	***Hours***
Force Continuum/ROE	2.00
Crowd Dynamics/Control/Formations	14.00
Communication Skills	2.00
MOOTW	2.00
Tactical Considerations	2.00
	22.00
NLW Systems	***Hours***
Oleoresin Capsicum	4.00
Empty Hand Control Techniques	19.00
Riot Baton Techniques	8.00
Nonlethal Munitions and Equipment	8.00
Portable Vehicle Barrier	5.00
66 mm Grenade Launcher	2.00
Modular Crowd Control Munition	2.00
	48.00
Nonlethal Weapons Instructor	***Hours***
Techniques of Military Instruction	10.00
NLW Scenario Exercise	10.00
	20.00
Subtotal =	**90.00**
Examinations =	***4.00***
Administration =	***6.00***
Total Hours =	***100.00***

4. Course Body

The course body consists of 15 subcourses. A quick synopsis of each subcourse is included below.

a. **Instructor Development**. This subcourse teaches and reinforces basic military instructor methodology. It identifies characteristics of good instructors and four ways they can improve their teaching ability. Additionally, it illustrates rules of conducting military instruction as well as the steps in planning a lesson and teaching classes.

b. **Force Continuum**. This subcourse introduces the five levels of resistance as well as identifying the levels of the force continuum. It illustrates influencing the use of force decisions and the definitions relating to deadly force. Lastly, it identifies the three preconditions for deadly force and seven authorized occasions for use of deadly force.

c. **Crowd Dynamics/Crowd Control**. This subcourse outlines the differences between crowds, mobs, and riots, and teaches the student basic crowd control techniques to apply in various situations. Upon completion, the student will be able to instruct others in both the behavioral aspects of a crowd during a civil disturbance and also various techniques of controlling a civil disturbance. It will identify factors that affect both individual and crowd behavior and it identifies all crowd control equipment needed, or essential, for mission accomplishment.

d. **Empty Hand Control Techniques**. This subcourse identifies safety factors to consider when instructing empty hand control techniques by illustrating the four Ms (mind, mass, movement, muscle) of self-defense, psychomotor skills, and phases of psychomotor skills. It illustrates levels of training and identifies common pressure points of the human body. It identifies hinges, appropriate angles of approach, and recommended patterns of movement. Instructors will learn blocking techniques, striking skills, control holds, escort positions, and restraint devices.

e. **Baton Techniques**. This subcourse identifies riot baton nomenclature and safety standards when conducting riot baton training. It identifies methods of teaching psychomotor skills and the phases of teaching psychomotor skills. It highlights baton-striking techniques, blocking techniques, and retention techniques.

f. **Military Operations Other Than War (MOOTW)**. This subcourse defines operations other than war and the difference between MOOTW and declarations of war. It helps identify the primacy of political objectives and the strategic aspects of MOOTW. It illustrates the range of operations and the principles of operations, as well as the types of military operations other than war.

g. **Rules of Engagement (ROE)**. This subcourse defines ROE and specific terms that apply to these rules. It also defines the purpose of ROE and provides the student an understanding of policies underlying the need for ROE. It identifies three sources for ROE and the two parts of ROE that will not change. It explains how ROE relate to using and applying NLW.

h. **Oleoresin Capsicum (OC) Training**. This subcourse explains how the three levels of OC are created, as well as the nomenclatures for all aerosol canisters. It highlights the different spray patterns, methods of delivery, and cautions concerning the traumatic hydraulic needle effect. It explains employment considerations and the three levels of OC contamination. Additionally, it explains the physical and mental effects of OC and trains students on different grips, draw methods, and proper stances when employing OC. It explains procedures for decontamination, recovery, and first aid.

i. **Types of Formations.** This subcourse explains the elements and individual responsibilities within a riot control formation. Additionally, it trains proper grip, balance, holding, and retention of the riot control shield. It highlights leadership requirements by teaching different commands and cadence for conducting riot control formations. It illustrates the different types of formations and the support elements of a riot control operation

j. **Tactical Considerations.** This subcourse addresses the preparations necessary for crowd control operations and the graduated response of a crowd control force. It identifies support and information required from host nation police and civil authorities. It describes the employment measures of military forces in crowd control and MOOTW environments. It highlights the actions, optional measures, and guidance for using NLW in crowd control and MOOTW environments.

k. **Nonlethal Munitions and Employment.** This subcourse identifies the types of nonlethal munitions and their intended purpose. It illustrates how nonlethal munitions fall within established "use of force" continuums. It identifies the types, characteristics, and ranges of all currently employed nonlethal munitions.

l. **Communications Skills.** This subcourse identifies the areas of nonverbal communication employed to determine an individual's state of mind. This, along with an understanding of the importance of personal space, gives the student the tools to resolve conflict with the amount of force as applied in the force continuum.

m. **Portable Vehicle Arrest Barrier.** This subcourse teaches the employment and maintenance of a force protection asset assigned to the nonlethal capability set. The PVAB is designed to stop a 7,500-pound vehicle going 45 mph. The PVAB protects gates and check points.

n. **Modular Crowd Control Munition.** This subcourse teaches the employment of the MCCM; the nonlethal version of the claymore mine. The MCCM is a force protection asset designed to help protect static defensive positions or check points.

o. **M-315 66-mm Grenade Launcher.** This subcourse teaches the installation, employment, and maintenance of the M-315. The M-315 increases the current maximum range of nonlethal sting ball munitions from approximately 50 meters to 100 meters. Additionally, this weapons platform can deploy flash bangs, smoke, and CS munitions.

Appendix D

Sample: Unit Training Schedules/Firing Tables

Nonlethal Individual Weapons Training Course Training Schedule: Unit Training Schedule Number 1.

Training Day #1					
Time	***Subject***	***Location***	***Uniform***	***Instr***	***Ref/Notes***
0800	Administration Time	Classroom	U1		
0830	Force Continuum	Classroom	U1		
1000	ROE	Classroom	U1		
1200	Chow	Mess Hall	U1		
1300	MOOTW	Classroom	U1		
1500	Communication Skills	Classroom	U1		
1700	Secure/COB				

Administration:

Uniforms:

U1: BDU/utility uniform
U2: BDU/utility uniform with LBE/harness and canteens
U3: BDU/utility uniform with LBE/harness, canteens, flak jacket, and helmet
U4: BDU/utility uniform with LBE/harness, canteens, flak jacket, helmet, and assigned weapon

Transportation:
Weapons:
Ammunition:
Required Equipment:
Screen
Video projector with central processing unit
Televisions
VCR
Miscellaneous:
All personnel are required to have a notebook, pencil, and pen.
Handouts prepared and passed out.

Training Day #2					
Time	***Subject***	***Location***	***Uniform***	***Instr***	***Ref/Notes***
0800	Nonlethal Equipment Brief	Classroom	U1		
0900	Pressure Point Control Techniques/ Open Hand Control Techniques	Classroom	U1		
1000	Pressure Point Control Techniques/ Open Hand Control Techniques (Practical Application)	PT Field	U2		
1200	Chow				
1300	Pressure Point Control Techniques/ Open Hand Control Techniques (Practical Application)	PT Field	U2		
1700	Debrief/Secure/COB				

Administration:

Uniforms:
U1: BDU/utility uniform
U2: BDU/utility uniform with LBE/harness and canteens
U3: BDU/utility uniform with LBE/harness, canteens, flak jacket, and helmet
U4: BDU/utility uniform with LBE/harness, canteens, flak jacket, helmet, and assigned weapon

Transportation:

Weapons:

Ammunition:

Required Equipment:
Screen
Video projector with central processing unit
Flexicuffs (one set of cuffs per two students)
Riot Training Bags (one bag per two students)

Miscellaneous:
All personnel are required to have a notebook, pencil, and pen.
Handouts prepared and passed out.
Medical personnel on station.

Training Day #3

Time	*Subject*	*Location*	*Uniform*	*Instr*	*Ref/Notes*
0800	Pressure Point Control	Classroom	U1		
	Techniques/Open Hand Control Techniques	Classroom	U1		
0900	Pressure Point Control Techniques/Open Hand Control Techniques (Practical Application)	PT Field	U1		
1200	Chow				
1300	Pressure Point Control Techniques/Open Hand Control Techniques (Practical Application)	PT Field	U2		
1700	Debrief/Secure/COB				

Administration:

Uniforms:
U1: BDU/utility uniform
U2: BDU/utility uniform with LBE/harness and canteens
U3: BDU/utility uniform with LBE/harness, canteens, flak jacket, and helmet
U4: BDU/utility uniform with LBE/harness, canteens, flak jacket, helmet, and assigned weapon

Transportation:

Weapons:

Ammunition:

Required Equipment:
Screen
Video projector with central processing unit
Flex cuffs (one set of cuffs per 2 students)
Riot Training Bags (one bag per 2 students)

Miscellaneous:
All personnel are required to have a notebook, pencil, and pen.
Handouts prepared and passed out.
Medical personnel on station.

Training Day #4					
Time	*Subject*	*Location*	*Uniform*	*Instr*	*Ref/Notes*
0800	Impact Weapons	Classroom	U1		
0900	Impact Weapons (Practical Application)	PT Field	U2		
1200	Chow				
1300	Impact Weapons (Practical Application)	PT Field	U2		
1700	Debrief/Secure/COB				

Administration:

Uniforms:

U1: BDU/utility uniform
U2: BDU/utility uniform with LBE/harness and canteens
U3: BDU/utility uniform with LBE/harness, canteens, flak jacket, and helmet
U4: BDU/utility uniform with LBE/harness, canteens, flak jacket, helmet, and assigned weapon

Transportation:

Weapons:

Baton (one baton per 2 students)
Training baton (one baton per 2 students)

Ammunition:

Required Equipment:

Screen
Video projector with central processing unit
Flex cuffs (one set of cuffs per 2 students)
Riot training bags (one bag per 2 students)
Three (3) riot training suits

Miscellaneous:

All personnel are required to have a notebook, pencil, and pen.
Handouts prepared and passed out.
Medical personnel on station.

Training Day #5					
Time	***Subject***	***Location***	***Uniform***	***Instr***	***Ref/Notes***
0800	Impact Weapons (Practical Application)	PT Field	U2		
1000	Crowd Dynamics/Crowd Control	Classroom	U1		
1130	Chow				
1230	Crowd Dynamics/Crowd Control	Classroom	U1		
1330	Riot Control Formations	Classroom	U1		
1430	Oleoresin Capsicum (OC) Aerosol Training	Classroom	U1		
1630	Debrief/Secure/COB				

Administration:

Uniforms:
U1: BDU/utility uniform
U2: BDU/utility uniform with LBE/harness and canteens
U3: BDU/utility uniform with LBE/harness, canteens, flak jacket, and helmet
U4: BDU/utility uniform with LBE/harness, canteens, flak jacket, helmet, and assigned weapon

Transportation:

Weapons:
Training baton (one baton per 2 students)
Baton (one baton per 2 students)

Ammunition:
Inert OC for classroom instruction

Required Equipment:
Riot training bags (one bag per 2 students)
Screen
Video projector with central processing unit
Three (3) riot training suits

Miscellaneous:
All personnel are required to have a notebook, pencil, and pen.
Handouts prepared and passed out.

Training Day #6					
Time	*Subject*	*Location*	*Uniform*	*Instr*	*Ref/Notes*
0800	OC Aerosol Training (Practical Application)	PT Field	U2		
1200	Chow				
1300	OC Aerosol Training (Practical Application)	PT Field	U2		
TBD	Debrief/Secure/COB				

Administration:

Uniforms:
U1: BDU/utility uniform
U2: BDU/utility uniform with LBE/harness and canteens
U3: BDU/utility uniform with LBE/harness, canteens, flak jacket, and helmet
U4: BDU/utility uniform with LBE/harness, canteens, flak jacket, helmet, and assigned weapon

Transportation:

Weapons:

Ammunition:
MK 4.inert OC (one per student)
MK 4 individual can OC (2 cans)

Required Equipment:
Five (5) large trash cans with liners
Two (2) water hoses with running water

Miscellaneous:
Medical personnel on station.

Training Day #7					
Time	*Subject*	*Location*	*Uniform*	*Instr*	*Ref/Notes*
0730	Movement to Range		U4		
0800	Nonlethal Munitions and Employment	Range Classroom	U1		
1030	Chow	Range			
1100	Safety Brief/Range Brief	Range	U4		
1130	Nonlethal Munitions (Live Fire)	Range	U4		
TBD	Evening Chow				
TBD	Safety Brief/Range Brief	Range	U4		
1800	Night Exercise Sun Set +30)	Range	U4		
TBD	Debrief/Clean Up	Range	U1		
TBD	Movement to Rear				
TBD	Secure/ COB				

Administration:

Uniforms:
U1: BDU/utility uniform
U2: BDU/utility uniform with LBE/harness and canteens
U3: BDU/utility uniform with LBE/harness, canteens, flak jacket, and helmet
U4: BDU/utility uniform with LBE/harness, canteens, flak jacket, helmet, and assigned weapon

Transportation: Yes

Weapons:
Assigned Weapons

Ammunition:
12-gauge rubber ball
12-gauge fin stabilized
12-gauge bean bag
12-gauge launching cup cartridge
40-mm foam baton
40-mm rubber ball
40-mm sponge grenade
Stingball grenades

Required Equipment:
Eight (8) 12-Gauge stinger launcher cup
Fifteen (15) targets
Eye and hearing protection
Fifteen (15) full-length riot shields
All students require munitions pouches, shin guards, face shields
Two (2) MRE per student
Video projector with central processing unit
Three (3) high-intensity light
Thirty (30) individual tactical lights

Miscellaneous:
Range safety officer.
Medical personnel on station.
Secondary means of communication (radios).
Conduct operational risk assessment.

Training Day #8					
Time	*Subject*	*Location*	*Uniform*	*Instr*	*Ref/Notes*
0700	Movement to Training Area #1		U2		
0730	Portable Vehicle Arresting Barrier (PVAB)	TA#1	U1		
0900	PVAB (Practical Application)	TA#1	U2		
1200	Chow				
1300	PVAB (Practical Application)	TA#1	U2		
1500	Debrief/Movement to Rear				
1530	Preventive Maintenance of PVAB				
1600	Secure/COB				

Administration:

Uniforms:
U1: BDU/utility uniform
U2: BDU/utility uniform with LBE/harness and canteens
U3: BDU/utility uniform with LBE/harness, canteens, flak jacket, and helmet
U4: BDU/utility uniform with LBE/harness, canteens, flak jacket, helmet, and assigned weapon

Transportation: Yes

Weapons:

Ammunition:

Required Equipment:
One (1) portable vehicle-arresting barrier (PVAB)
Video projector with central processing unit

Miscellaneous:
(1) MRE per student.

Training Day #9					
Time	*Subject*	*Location*	*Uniform*	*Instr*	*Ref/Notes*
0730	Movement to Range		U4		
0800	Modular Crowd Control Munition-Ground Emplaced (MCCM-GE)	Range Classroom	U1		
0900	Safety Brief/Range Brief	Range	U4		
0930	MCCM-GE (Practical Application	Range	U4		
1030	Chow				
1130	Vehicle Mounted Nonlethal Grenade Launcher (VMNLGL)	Range Classroom	U4		
1300	VMNGL	Range	U4		
1730	Debrief/Movement to Rear				
1800	Secure/COB				

Administration:

Uniforms:

U1: BDU/utility uniform
U2: BDU/utility uniform with LBE/harness and canteens
U3: BDU/utility uniform with LBE/harness, canteens, flak jacket, and helmet
U4: BDU/utility uniform with LBE/harness, canteens, flak jacket, helmet, and assigned weapon

Transportation: Yes

Weapons:

Vehicle Mounted Nonlethal Grenade Launcher (VMNLGL)

Ammunition:

(2) MCCM-GE, (2) 66-mm Riot Control Training Grenade (L97E1)

Required Equipment:

Eye and ear protection
Video projector with central processing unit

Miscellaneous:

Range safety officer.
Medical personnel on station.
Secondary means of communication.
Two (2) MRE per student.

Training Day #10					
Time	*Subject*	*Location*	*Uniform*	*Instr*	*Ref/Notes*
0700	Movement Down Range				U4
0730	Tactical Considerations	MOUT Classroom		U1	
0930	Safety Brief/Range Brief	MOUT Facility		U4	
1000	Tactical Exercise* (Practical Application)	MOUT Facility		U4	
1200	Chow				
1300	Tactical Exercise (Practical Application)	MOUT Facility		U4	
1700	Debrief/Movement to Rear				
1730	Weapons Maintenance				
1800	Secure/COB				

Administration:

Uniforms:
U1: BDU/utility uniform
U2: BDU/utility uniform with LBE/harness and canteens
U3: BDU/utility uniform with LBE/harness, canteens, flak jacket, and helmet
U4: BDU/utility uniform with LBE/harness, canteens, flak jacket, helmet, and assigned weapon

Transportation: Yes

Weapons:
Assigned weapon
Inert OC individual (one per student)
Training batons (one per student)
Fifty (50) practice grenades

Ammunition:

Required Equipment:
Three (3) riot training suit
All students require munitions pouches, flex cuffs, shin guards, face shields
Video projector with central processing unit
Fifteen (15) full length riot shields

Miscellaneous:
Range safety officer.
Medical personnel present.
Secondary means of communication.
(2) MRE's per student.

UNITED STATES MARINE CORPS
Marine Corps Detachment
Fort Leonard Wood, MO 65473

1.	LESSON DESIGNATOR:	Interservice Nonlethal Individual Weapons Instructor Course (INWIC). 13A (INIWIC.13A)
2.	LESSON TITLE:	Scenario Practical Exercise
3.	SCHOOL/COURSE:	Nonlethal Individual Weapons Instructor Course
4.	STUDY ASSIGNMENT:	Review course curriculum
5.	ESTIMATED TIME:	8 hours
6.	INSTRUCTORS REQUIRED:	(1) primary, (1) alternate per ratio: 1:10
7.	INSTRUCTOR REFERENCES:	INIWIC POI
8.	DATE PREPARED:	September 2001
9.	PRESENTATION METHOD:	Practical Exercise
10.	TRAINING AIDS:	Warning Order and five paragraph operations order for exercise, rubber rifles/shotguns, batons, shields, Inert MK4 OC canister, bullhorn, radio
11.	STUDENT EQUIPMENT:	(1) 782 gear (1) Kevlar with face shield (1) Shin guards

5. Unit Training Schedule Number 2

Table D-1. NLW Training Course 5-Day Training Schedule

Day/date	*Time*	*Event*	*Location*	*Instructor*	*Notes*
Mon/07 Jan 02	0730-0745	Welcome Aboard	Bldg RR202	MSgt Dixon	
T-1	0745-0845	Force Continuum	SOTG	TBD	L
	0900-1015	MOAB	SOTG	TBD	L
	1030-1130	OC	SOTG	TBD	L
	1130-1300	Noon Meal	Mess Hall	Company GySgt	
	1300-1400	OC Practical Application	SOTG	TBD	PA
	1400-1700	OC Certification	SOTG	TBD	PA
Tues/08 Jan 02	0730-0830	ROE	SOTG	TBD	L
T-2	0845-0945	Crowd Control	SOTG	TBD	L
	1000-1130	MDTS	SOTG	TBD	L, PA
	1130-1300	Noon Meal	Mess Hall	Company GySgt	
	1300-1700	MDTS	SOTG	TBD	L, PA
Wed/09 Jan 02	0730-1130	Batons	SOTG	TBD	L, PA
T-3	1130-1300	Noon Meal	Mess Hall	Company GySgt	
	1300-1500	Batons	SOTG	TBD	L, PA
	1500-1700	Formations	SOTG	TBD	L
Thurs/10 Jan 02	0730-1130	Formations	SOTG	TBD	PA
T-4	1130-1300	Noon Meal	Mess Hall	Company GySgt	
	1300-1700	Nonlethal Munitions/shotguns/ M203/sustainment training	Square bay/M203	TBD	L, PA
Fri/11 Jan 02	0730-1130	Nonlethal Munitions/stinger ball grenades/shotgun/M203	Square Bay/M203	TBD	PA
T-5	1130-1300	Noon Meal	Mess Hall	Company GySgt	
	1300-1700	Nonlethal Munitions/stinger ball grenades/shotgun/M203	Nonlethal Range 1&2	TBD	PA

Notes:

L = Lecture
PA = Practical Application

MDTS = Manadnock Defensive Tactics System
MOAB = Management of Aggressive Behavior
SOTG = Special Operations Training Group (USMC training unit)

6. Sample Unit Training Schedule Number 3

Table D-2. NLW Training Course 10-Day Training Schedule

DAY/DATE	***TIME***	***EVENT***	***LOCATION***	***INSTRUCTOR***	***NOTES***
Mon/29 Oct 01	0730-0745	Welcome Aboard	Bldg RR-202	MSgt Dixon	
T-1	0745-0845	Force Continuum	PT Field	Sgt Whitworth	L
	0900-1015	MOAB	PT Field	GySgt German	L
	1030-1130	OC	PT Field	Sgt Hoskins	L
	1130-1300	Noon Meal	Mess Hall	Company Gunny	
	1300-1400	OC Practical Application	PT Field	Sgt Hoskins	PA
	1400-1700	OC Certification	PT Field	Sgt Hoskins	PA
Tues/30 Oct 01	0730-0830	ROE	PT Field	Cpl Fewox	L
T-2	0845-0945	Crowd Control	PT Field	SSgt Parrish	L
	1000-1130	MDTS	PT Field	Sgt Johnson	L, PA
	1130-1300	Noon Meal	Mess Hall	Company Gunny	
	1300-1700	MDTS	PT Field	Sgt Johnson	L, PA
Wed/31 Oct 01	0730-1130	Batons	PT Field	Sgt Johnson	L, PA
T-3	1130-1300	Noon Meal	Mess Hall	Company Gunny	
	1300-1500	Batons	PT Field	Sgt Hoskins	L, PA
	1500-1700	Formations	PT Field	GySgt German	L
Thur/1 Nov 01	0730-1130	Formations	PT Field	Sgt Whitworth	PA
T-4	1130-1300	Noon Meal	Mess Hall	Company Gunny	
	1300-1700	Nonlethal Munitions	Nonlethal Range 1&2	GySgt German	L, PA
Fri/2 Nov 01	0730-1130	Nonlethal Munitions	Nonlethal Range 1&2	SSgt Parrish	PA
T-5	1130-1300	Noon Meal	Mess Hall	Company Gunny	
	1300-1700	Nonlethal Munitions	Nonlethal Range 1&2	SSgt Parrish	PA
Mon/5 Nov 01	0730-1130	Nonlethal Grenades Formation Fire	Nonlethal Range 1&2	GySgt German	L, PA
T-6	1130-1300	Noon Meal	Mess Hall	Company Gunny	
	1130-1700	Formation Fire/Sustainment Training	Nonlethal Range 1&2	SSgt Blackler	PA
Tues/6 Nov 01	0730-0830	Introduction to MOOTW	PT Field	Maj Desy	L
T-7	0830-1130	Vehicle Checkpoints/Vehicle Searches	PT Field	Sgt Whitworth	L, PA
	1130-1300	Noon Meal	Mess Hall	Company Gunny	
	1300-1500	Vehicle	PT Field	SSgt Parrish	PA

Table D-2. NLW Training Course 10-Day Training Schedule

DAY/DATE	TIME	EVENT	LOCATION	INSTRUCTOR	NOTES
		Checkpoints/Vehicle Searches			
	1500-1700	Roadblocks/Personnel Searches	PT Field	SSgt Blackler	L, PA
Wed/7 Nov 01	0730-1130	Roadblock/Personnel Searches	PT Field	SSgt Blackler	PA
T-8	1130-1300	Noon Meal	Mess Hall	Company Gunny	
	1300	Operations Order	Nonlethal Classroom	MSgt Dixon	Company Staff
Thurs/8 Nov 01	0730-1000	Confirmation Brief	PT Field	Maj Desy	
T-9	1000-1200	Insert	Combat Town	SSgt Parrish	
	1200-0000	STX	Combat Town	SSgt Parrish	
Fri/9 Nov 01	0001-1200	STX	Combat Town	SSgt Parrish	
T-10	1200-1300	Cleanup/Debrief	Combat Town	Company Gunny	
	1300	Graduation	Combat Town	SOTG OIC	

Notes:

L = Lecture

PA = Practical Application

MOAB = Management of Aggressive Behavior

MDTS = Manadnock Defensive Tactics System

STX = Situational Training Exercise

Appendix E

NLW Capability Sets

This appendix provides the current NLW capability sets for the services. These sets continue to be upgraded and changed according to service needs and the availability of new technologies. The U.S. Army and USMC capability sets are standard for the service, while the USAF, USN, and USCG are still developing and fielding their sets.

1. U.S. Army NLW Capability Set

Table E-1 U.S. Army Nonlethal Capability Set (Version FY 2001)

	Name	*Quantity*	*Unit Cost*	*Item Total*
GSAPMDK5X250AFHC	Non-Ballistic Riot Face Shields	200	$42.50	$8,500.00
GSAA TIPMBS9	Non-Ballistic Riot Body Shields Paulson	150	$88.50	$13,275.00
GSAATIUSMCSG	Non-Ballistic Riot Shin Guards	150	$60.00	$9,000.00
PT702MTUSA	Ballistic Face Shields	10	$246.00	$2,460.00
PT2237	Ballistic Body Shields	10	$1,052.00	$10,520.00
GSASF6SL	Sure Fire Field Light	10	$185.82	$1,858.20
MON2800 w/BCS-1	36" Hickory Riot Baton with Grommet	200	$15.50	$3,100.00
EADBH	Eagle Duty Baton Holder OD	200	$5.50	$1,100.00
GSAATIA2M	Portable Bullhorn	10	$82.00	$820.00
AAPB35W1	Ground Mount Bullhorn	8	$815.00	$6,520.00
AAPBWH10001	Megavox Wireless MIC	8	$0.00	$0.00
AASS450	Megavox Stand	8	$0.00	$0.00
GSASF6P	Individual High Intensity Light	15	$42.74	$641.10
GSAATIEATAVLFP6POD	Eagle Sure Fire 6P Holster	15	$8.75	$131.25
GSASFP60	Individual High Intensity Light Accessory Kit	15	$12.53	$187.95
XENNCSUSA	High Intensity Light	8	$2,795.00	$22,360.00
NIKFLEXCUFF	Disposable Forearm/Ankle Cuffs (Single)	1000	$0.70	$700.00
NIKCUTTER	Flex Cuff Cutters	2	$9.00	$18.00
GSAATIEAMK4	MK-4 Pouch OD	200	$6.75	$1,350.00
GSADEF5149	Inert Individual Training (OC) Dispenser	200	$6.07	$1,214.00
GSASF622	12-Gauge Shotgun High Intensity Light	50	$139.65	$6,982.50
GSAATIEAUPMS	12-Gauge Utility (25-Round) Pouch	50	$23.65	$1,182.50
GSAATISSSCQTR	12-Gauge Buttcuff Carrier (6-Round)	50	$10.75	$537.50
GSAATISASDD60D	Diversionary/Sting Ball Grenade Pouch	25	$47.25	$1,181.25
GSAATIEA40MM6	40-mm Carrying Pouch	100	$13.65	$$1,365.00
GSAATIRSC3	Caltrop	250	$10.50	$2,625.00
GSAATIFS333	Riot Training Suit with Accessories	4	$638.50	$2,554.00
GSAMONUTB	Riot Training Bag	10	$80.00	$800.00
BNCCELLS	C-Cell Batteries	0	$1.00	$0.00
BNAAACELLS	AA Batteries	0	$0.75	$0.00
GSASFDL123A	3V Lithium Batteries	0	$2.85	$0.00
BN9V	9V Batteries	0	$2.00	$0.00
ATIPMI	Packaging Marking and Integration Fee			$20,300.00
	Hardware Tools		**Item Total**	**$125,683.25**
	U.S. Army NLW Capability Set	**5**	**$125,683.25**	**$628,416.25**

2. USMC Nonlethal Weapons Capability Set

a. USMC NLW Company-Size Set

Table E-2 U.S. Marine Corps FY02 Company-Size Nonlethal Capability Set

TAMCN C41202E	*NSN 9999-01-464-2285*	*SET A (Company Size)*		
NSN/Product ID	***Name***	***Quantity***	***Unit cost***	***Item Total***
GSAATINCSPMDK5	Non-Ballistic USMC Riot Face Shields (S/R)	200	$42.50	$8,500.00
GSAATIPMBS9	Non-Ballistic Riot Body Shields Paulson	60	$88.50	$5,310.00
GSAATIUSMCSG	Non-Ballistic Riot Shin Guards	200	$60.00	$12,000.00
GSAATI23/36	USMC 23"/36" Riot Baton (Expandable) (w/Holster)	200	$73.25	$14,650.00
GSAATITRAINING BATON	USMC Riot Baton (Practice) (Price TBD)	200	$20.00	$4,000.00
GSAATIA2M	Portable Bullhorn	12	$82.00	$984.00
GSAATIDC/W	USMC Disposable Handcuffs	400	$0.94	$376.00
CGAATIDC/T	USMC Disposable Handcuff Training Device	100	$0.94	$94.00
GSAATIEAMK4	MK-4 Pouch OD	200	$6.75	$1,350.00
GSADEF5049	MK-4 Individual Live (OC) Dispenser (Pepper Spray)	200	$7.29	$1,458.00
GSADEF5099	MK-9 Live Squad (OC) Dispenser (Pepper Spray)	40	$26.20	$1,048.00
GSATIEAMK9	MK-9 Pouch OD	40	$11.00	$440.00
GSAATIDEF5846L	MK-46 Live (OC) (Horizontal) (Pepper Spray)	12	$300.00	$3,600.00
GSAATIDEF1515	MK-46 Sling	12	$30.25	$363.00
GSAATIDEF5520	MK-46 Hardware Kit	4	$68.50	$274.00
GSAATIDEF5546	MK-46 Refill Unit (OC) Live (Pepper Spray)	4	$46.25	$185.00
GSADEF5149	MK-4 Inert Individual Training (OC) Dispenser	200	$6.07	$1,214.00
GSADEF5199	MK-9 Inert Squad Training (OC) Dispenser	40	$20.10	$804.00
GSAATIDEF5946L	MK-46 Inert Training (OC) (Horizontal) (w/Dye)	12	$240.00	$2,880.00
GSADEF5546I	MK-46 Refill Unit (OC) Inert Training (w/Dye)	4	$27.50	$110.00
GSAATIFS333	Riot Training Suit w/Accessories (F.I.S.T.)	12	$638.50	$7,662.00
GSAMONUTB	Riot Training Strike Bag	24	$80.00	$1,920.00
GSAATIRSC3	Caltrop	400	$10.50	$4,200.00
TBD	Stinger Road Side Spike Kit (15.5') (Price TBD)	12		
XENNLCSUSMC	High Intensity Light (Xenonics)	6	$2,975.00	$17,850.00
GSAATIEATAVLFP6POD	Eagle Sure Fire 6P Holster	60	$8.75	$525.00
GSASF6P	Individual High Intensity Light	60	$42.74	$2,564.40
GSASFP60	Individual High Intensity Light Accessory Kit	30	$12.53	$375.90
GSASFDL123A	3V Lithium Batteries	120	$2.85	$342.00
GSAATISSSCQTR	12-Gauge Buttstock Cuff	36	$10.75	$387.00
GSAATIEAUPMS	12-Gauge Utility (25-Round) Pouch	72	$23.65	$1,702.80
GSAATIEA40MM6	40-mm Carrying Pouch	36	$13.65	$491.40
GSAATISASDD60D	Diversionary/Sting Ball Grenade Pouch	60	$47.25	$2,835.00
C3I9512	12-Gauge Launching Cup (Availability and Price TBD)	36		
1005-01-371-5639/ 1005-01-251-8578	12-Gauge Mossberg 500A2 / 590 Shotgun	36	$250.00	$9,000.00
ATIPMI	Packaging Marking and Integration Fee			$11,578.00
	Hardware Totals		**Item Total**	**$121,073.50**

b. USMC NLW Platoon-Size Set

Table E-3 USMC FY02 NLW Capability Set—Platoon Size

TAMCN C41202E	*NSN 9999-01-464-2285*	*SET B (Platoon Size)*		
NSN/Product ID	***NAME***	***Quantity***	***Unit Cost***	***Item Total***
GSAATINCSPMDK5	**Non-Ballistic USMC Riot Face Shields (S/R)	50	$42.50	$2,125.00
GSAATIPMBS9	Non-Ballistic Riot Body Shields Paulson	15	$88.50	$1,327.50
GSAATIUSMCSG	Non-Ballistic Riot Shin Guards	50	$60.00	$3,000.00
GSAATI23/36	USMC 23"/36" Riot Baton (Expandable) (w/Holster)	50	$73.25	$3,662.50
GSAATITRAINING BATON	USMC Riot Baton (Practice) (Price TBD)	50	$20.00	$1,000.00
GSAATIA2M	Portable Bullhorn	3	$82.00	$246.00
GSAATIDC/W	USMC Disposable Handcuffs	100	$0.94	$94.00
GSAATIDC/T	USMC Disposable Handcuff Training Device	25	$0.94	$23.50
GSAATIEAMK4	MK-4 Pouch OD	50	$6.75	$337.50
GSADEF5049	MK-4 Individual Live (OC) Dispenser (Pepper Spray)	50	$7.29	$364.50
GSADEF5099	MK-9 Live Squad (OC) Dispenser (Pepper Spray)	10	$26.20	$262.00
GSATIEAMK9	MK-9 Pouch OD	10	$11.00	$110.00
GSAATIDEF5846L	MK-46 Live (OC) (Horizontal) (Pepper Spray)	3	$300.00	$900.00
GSAATIDEF1515	MK-46 Sling	3	$30.25	$90.75
GSAATIDEF5520	MK-46 Hardware Kit	2	$68.50	$137.00
GSAATIDEF5546	MK-46 Refill Unit (OC) Live (Pepper Spray)	2	$46.25	$92.50
GSADEF5149	MK-4 Inert Individual Training (OC) Dispenser	50	$6.07	$303.50
GSADEF5199	MK-9 Inert Squad Training (OC) Dispenser	10	$20.10	$201.00
GSAATIDEF5946L	MK-46 Inert Training (OC) (Horizontal) (w/Dye)	3	$240.00	$720.00
GSADEF5546I	MK-46 Refill Unit (OC) Inert Training (W/Dye)	2	$27.50	$55.00
GSAATIFS333	Riot Training Suit W/Accessories (F.I.S.T.)	3	$638.50	$1,915.50
GSAMONUTB	Riot Training Strike Bag	3	$80.00	$240.00
GSAATIRSC3	Caltrop	50	$10.50	$525.00
TBD	Stinger Road Side Spike Kit (15.5') (Price TBD)	3		
XENNLCSUSMC	High Intensity Light (Xenonics)	4	$2,975.00	$11,900.00
GSAATIEATAVLFP6POD	Eagle Sure Fire 6P Holster	50	$8.75	$437.50
GSASF6P	Individual High Intensity Light	50	$42.74	$2,137.00
GSASFP60	Individual High Intensity Light Accessory Kit	25	$12.53	$313.25
GSASFDL123A	3V Lithium Batteries	100	$2.85	$285.00
GSAATISSSCQTR	12-Gauge Buttstock Cuff	15	$10.75	$161.25
GSAATIEAUPMS	12-Gauge Utility (25 Round) Pouch	30	$23.65	$709.50
GSAATIEA40MM6	40-mm Carrying Pouch	50	$13.65	$682.50
GSAATISASDD60D	Diversionary/Sting Ball Grenade Pouch	50	$47.25	$2,362.50
CSI9512	12-Gauge Launching Cup (Availability and Price TBD)	15		
1005-01-371-5639/ 1005-01-251-8578	12-Gauge Mossberg 500A2 / 590 Shotgun	15	$250.00	$3,750.00
ATIPMI	Packaging Marking and Integration Fee			$3,000.00
	Hardware Totals		**Item Total**	**$43,471.25**

** Riot shield. Marine Corps security forces are currently evaluating to use in conjunction with each other as listed below. They are not currently deployed but are undergoing field evaluation.
(1) The six feet Armadillo Interlocking Shield for the hard baseline.
(2) The four feet.. Armadillo Interlocking Shield for the soft baseline.

3. USN NLW Capability Set

a. The USN is in the process of developing NLW capabilities and capability sets.[*] The sets will support the specific areas of operation, mission, and/or user.

b. The first version will address the force protection mission and will be based on other service-approved items and some longer-range commercial-off-the-shelf items currently being upgraded for military use. The mobile security forces, master at arms, and shipboard force protection units are the targeted users.

c. A notional capability set for the Mobile Security Forces is listed below.

Table E-4. Notional USN Capability Set

Name	*Nomenclature (Mfr Representative)*	*Quantity*	*Unit Cost*	*Case Size*	*Cost Extended*
Fin Stabilize Round		1 Case	$5	100	$400
Stingshot		1 Case	$5	100	$400
Shotbag		1 Case	$5	100	$450
Flamethrower Round		1 Case	$8	100	$750
Tactical Blast Bursting Round		1 Case	$7	100	$650
Ultraflash Bursting Round		1 Case	$8	100	$750
Launchable Stun Munition	POP, Inc. T-444	1 Case	$60	24	$1,440
40-mm Sparkshot		1 Case	$35	24	$840
40-mm Stingshot		1 Case	$25	24	$600
40-mm Shotbag		1 Case	$25	24	$600
40-mm OC		1 Case	$28	24	$660
Stinger	POP, Inc. I-452	1 Case	$60	24	$1,440
Stun	POP, Inc. T-470	1 Case	$55	24	$1,320
Flash/Bang Multi	POP, Inc. D-451	1 Case	$65	24	$1,560
OC	POP, Inc. I-452	1 Case	$65	24	$1,560
12-Gauge Launching Blanks		200	$3		$600
Grenade Training Rounds		12	$65		$780
12-Gauge Training Rounds		24	$3		$72
40-mm Training Rounds		12	$20		$240
12-Gauge Shotguns	Standard Issue/Modified	12	$1,000		$12,000
12-Gauge Launching Adapter		6	$325		$1,950
Face Shields		15	$100		$1,500
Helmets		15	$200		$3,000

---

[*] In FY 01, the USN fielded an USMC capability set to Naval Air Station, Roosevelt Roads, Puerto Rico.

Table E-4. Notional USN Capability Set

Name	Nomenclature (Mfr. Representative)	Quantity	Unit Cost	Case Size	Cost Extended
Tactical Vest		15	$300		$4,500
Bullhorn		2	$100		$200
Medical Kit		2	$50		$100
Searchlights		2	$100		$200
Field Pack		15	$100		$1,500
Holsters/Belts		15	$50		$750
Ordnance Packaging, Handling, Storage and Transportation (PHS&T)					$1,500
Misc. (Certification, Training and Documentation)					$3,000
			Total Cost Basic Kit		**$45,312**

Table E-5. Notional USN Capability Set

Name	Nomenclature (Mfr. Representative)	Quantity	Unit Cost	Case Size	Cost Extended
Stingmore Mine (Std Body) with Rubber Ball Projectiles	POP, Inc. I-454	24	$350		$8,400
Stingmore Mine Stand		6	$100		$600
Electric Firing Boxes		8	$35		$280
Area Lighting					$1,000
Electric Generator					$500
Signs and Placards					$500
Personal NVGs and Opticals		6	$2,000		$12,000
Additional PHS&T					$2,000
Misc. (Certification, Training and Documentation)					$2,000
Total Cost Augmented Basic Kit					**$27,280**
Total Cost (Basic plus Augment)					$72,592
Portable Vehicle Arresting Barrier	U.S. Army	1	$40,000		$40,000
Running Gear Entanglement System (RGES)	NSWCDD	1000 ft system	$25,000		$25,000
Antimaterial Round (USCG)		24	$12		$288
Spike Strips		2	TBD		
Improved Sensor (s)					
RF			TBD		

Table E-5. Notional USN Capability Set

Name	Nomenclature (Mfr. Representative)	Quantity	Unit Cost	Case Size	Cost Extended
IR			TBD		
Acoustic			TBD		
Chemical Detection	SABRE 2000	1	$20,000		$20,000
Ordnance Packaging, Handling, Storage and Transportation (PHS&T)					$6,000
Misc. (Certification, Training and Documentation)					$2,000
Total Cost Vehicle Supplement					**$93,288**
Total Cost Basic, Augmented, and Vehicle Supplement					**$165,880**

4. USAF NLW Capability Set

Table E-6. USAF NLW Capability Set

NSN	Item Description	EA	Item $	Item Total
GSADEF5149	Inert Individual (OC) Training Dispenser	1	$6.07	$6.07
GSADEF5846H	Mid-Size Riot Control Dispenser (Chemical)	3	$318.75	$956.25
GSADEF5546	Riot Control Dispenser (OC) Agent Quarts	1	$46.25	$46.25
	12-Gauge Shotgun (Area CDC/Rubber Pellet)	36	$4.46	$160.56
	12-Gauge Shotgun (Point/Fin Stabilized)	36	$2.85	$102.60
CSI (USMC) GG04	Stun Hand (Sting Ball) Grenade	50	$35.00	$1,750.00
M84	Stun Hand Grenade/Flash-Bang Device	50	$78.00	$3,900.00
GSAPMDK5X/AF	Non-Ballistic Face Shields	13	$29.50	$383.50
GSAPMBS9	Non-Ballistic Full Length Riot Shield	13	$88.50	$1,150.50
GSAUSMCSG	Non-Ballistic Shin guards	13	$6.000	$780.00
GSAATI23/36	36" Riot Baton w/Holster	13	$73.25	$952.25
GSAM2	Portable Megaphone	1	$82.00	$82.00
EAAFCUSTOM	Tactical Vest Rough Estimate	3	$150.00	$450.00
HARG700	Riot Control Gloves	13	$52.20	$678.60
Champro	Protective Cups	13	$7.25	$94.25
	40-mm (Point) Sponge Round	54	$23.00	$1,242.00
	40-mm (Area) Crowd Dispersal Cartridge Grenade	54	$22.00	$1,188.00
GSAF333	Riot Baton Training Suit	1	$638.50	$638.50

Table E-6. USAF NLW Capability Set

NSN	***Item Description***	***EA***	***Item $***	***Item Total***
MONDC/W	Disposable Forearm/Ankle Cuffs (FLEX)	40	$0.94	$37.60
GSADEF5049/ EA	Individual (OC) Dispenser with Carrier	13	$7.29	$94.77
	12-Gauge Utility Pouch	2	$24.00	$48.00
	12-Gauge Buttstock Cuff	3	$11.00	$33.00
	40-mm Carrying Pouch	2	$14.00	$28.00
	Training Strike Bag	2	$80.00	$160.00
	Cost of Clip-Lok Boxes			$480.00
	Approximate freight shipped via commercial freight into CONUS			$1,500.00
			Total Estimate	**$16,942.70**
Proposal to add TASERS to USAF Capability Set after testing.				

5. Notional USCG NLW Capability Set

Table E-7. Notional USCG NLW Capability Set

Reference: (a) COMDTINST M16247.1B Maritime Law Enforcement Manual (MLEM) (b) Project Ericson Website	
Expandable Baton	
Restraint Devices	Hand Cuffs, Flex Cuffs, Leg Irons, Migrant Restraint Harness
Individual OC Spray	
Non-Ballistic Riot Face Shield	
Non-Ballistic Riot Body Shield	
Non-Ballistic Riot Shin Guard	
Running Gear Entanglement Systems	

The following information describes each of the components of the NLW capability set currently deployed by the USCG. (The USCG has not designated a standard set of nonlethal gear. The using unit determines the gear needs.)

6. Restraint Devices

a. Handcuffs are the preferred restraint devices when dealing with a small number of subjects. Handcuffs should be steel, chain-linked construction capable of being double locked, with nickel, blued, or stainless steel finish.

b. Using flexible handcuffs for the temporary restraint of subjects is also authorized. Since they do not require a key and are relatively easy to apply, flexible handcuffs are particularly useful and appropriate when dealing with large numbers of subjects (e.g. migrants).

c. Leg irons are authorized restraint devices and may be a preferred option for restraining subjects for extended periods of time.

d. Migrant restraint harnesses are authorized when dealing with large numbers of subjects.

7. Expandable Baton

The expandable baton is the only intermediate weapon authorized for USCG personnel to carry. It should have a steel shaft that is silver in color and corrosion resistant (e.g. chrome, nickel, stainless steel) and have no knurling or sharp edges. It should have a foam or similar, non-slip grip. It should be rigid when deployed and must be no less than 15 inches and no greater than 22 inches when expanded. Spring loaded or other trigger style batons are not authorized.

8. OC Pepper Spray

Only chemical irritants containing the active ingredient OC are authorized for USCG personnel to carry. The aerosol must contain no greater than 10 percent and no less than 5 percent OC concentration. It shall have a water-based carrier, be nonflammable, nontoxic and produce a straight stream delivery (not fog).

9. Helmet Riot Face Shield

Comparable to DOD type. Not currently deployed. Undergoing field evaluation.

10. Riot Shield

Two shields are currently being evaluated. A 24 x 48 inches clear lexan weighing 14 pounds and a smaller version measuring 20 x 36 inches. They are not currently deployed. Undergoing field evaluation.

REFERENCES

Joint

JP 1-02, *DOD Dictionary of Military and Associated Terms*, 12 Apr. 2001

JP 3-07, *Joint Doctrine for Military Operations Other Than War*, 16 Jun. 1995

JP 3-07.3, *JTTP for Peace Operations*, 12 Feb.1999

JP 3-57, *Doctrine for Joint Civil-Military Operations*, 8 Feb. 2001

JP 3-58, *Joint Doctrine for Military Deception*, 31 May 1996

JP 3-61, *Doctrine for Public Affairs in Joint Operations*, 14 May 1997

JP 4-0, *Joint Doctrine for Logistic Support of Joint Operations*, 6 Apr. 2000

Multi-Service

FM 3-100.12, MCRP 5-12.1C, NTTP 5-03.5, AFTTP(I) 3-2.34, *Risk Management* for *Multiservice Tactics, Techniques, and Procedures*, 15 Feb. 2001

FM 3-31.1, MCWP 3-36, *Army and Marine Corps Integration in Joint Operations*, 21 Nov. 2001

Army

AR 190-14, *Military Police Carrying of Firearms and Use of Force for Law Enforcement and Security Duties*, 12 Mar. 1993

AR 350-38, *Training Devices Policies and Management*, 15 Oct. 1993

AR 27-1, *Judge Advocate Legal Services*, 30 Sep. 1996

FM 3-0, *Operations*, 14 Jun. 2001

FM 3-05.30 (33-1), *Psychological Operations*, 19 Jun. 2000

FM 19-15, *Civil Disturbances*, 25 Nov. 1985

FM 27-10, *The Law of Land Warfare*, 18 Jul. 1956

FM 100-19, *Domestic Support Operations*, 1 Jul. 1993

TC 90-1, *Training for Urban Operations*, 1 Apr. 2002

TRADOC Pamphlet 525-5, *Force XXI Operations, A Concept for the Evolution of Full Dimensional Operations for the Strategic Army of the Twenty-First Century*, 1 Aug. 1994

TRADOC Pam 350-9, *TRADOC Training Devices for Army wide Use*, 30 Sep. 1993

TRADOC Regulation 350-70, *Systems Approach to Training Management, Processes, and Products*, 9 Mar. 1999.

TRADOC Regulation 385-2, chg.1, TRADOC Safety Program, 10 Oct. 2000.

Marine Corps

MCWP 3-33.4, *Domestic Support Operations*, Jul. 1993

I MEF FMF, After Action Report—UNITED SHIELD, 7 Jun. 1995

MCO 1510.112, *Individual Training Standards for Nonlethal Weapons*, 27 May 1998

MCO 5500.6F, *Arming of Security and Law Enforcement Personnel and the Use of Force*, 20 Jul. 1995

MCRP 5-12.1A, *The Law of Land Warfare.*

MCO 1630.4A, *Law Enforcement and Physical Security Activities*, 19 Jun. 1991.

MCO 1510.114, *Individual Training Standards (ITS) for Antiterrorism Force Protection (AT/FT) System*, 30 Jan. 1998.

MCWP 3-11.1, *Marine Rifle Company*, Feb. 1978

MCRP 3-11.1A, *Commander's Tactical Handbook*, 30 Jul. 1988. .

MCWP 3-11.2 *Marine Rifle Squad*

Marine Corps Gazette, *"Less-Lethal" Force in Operation UNITED SHIELD*, Sep. 1995.

Navy

Naval Doctrine Publication 1, Naval Warfare (U), 28 Mar. 1994

NWP 1-14M, *The Commander's Handbook on the Law of Naval Operations*, Oct. 1995

SECNAVINST 5500.29B, *Use of Deadly Force and the Carrying of Firearms by Personnel of the Department of the Navy in Conjunction with Law Enforcement Security Duties and Personal Protection*, 29 Sep. 1992

Air Force

AFI 31-201, *Security Police Standards and Procedures*, 4 Dec. 2001

AFI 31-207, *Arming and Use of Force by Air Force Personnel*, 1 Sep. 1999

AFMAN 31-201V6, *Civil Disturbance*, 17 May 2002

AFMAN 31-222, *Security Forces Use of Force Manual*, 1 Jun. 2001

Coast Guard

COMDTINST M16247.1B, Maritime Law Enforcement Manual, 9 Aug. 2001

Other

DODD 3000.3-R, *Policy for Nonlethal Weapons*, 9 Jul. 1996

DODD 3025.1, *Military Support to Civil Authorities (MSCA)*, 15 Jan. 1993

DODD 5100.77, *DOD Law of War Program*, 9 Dec. 1998

DODD 3025, *I-M Manual for Civil Emergencies*, 2 Jun. 1994

DODD 3025.12, *Military Assistance for Civil Disturbances*, 4 Feb. 1994

DODD 3025.13, *DOD Resources in Support of US Secret Service,* 13 Sep. 1985

DODI 5000.2, *Operation of the Defense Acquisition System*, 5 Apr. 2002

DODR 5000.1, *Defense Acquisition*, 23 Oct. 2000, Ch 1, 4 Jan. 2001

DOD Joint Concept for Nonlethal Weapons, 5 Jan. 1998

DODD 3025.12, *Military Assistance for Civil Authorities*, 18 Feb. 1997

DODD 3025.16, *Military Assistance for Civil Authorities Liaison Office (EPLO) Program*, 18 Dec. 2000

DODI 5000.2, *Operation of the Defense Acquisition System*, 5 Apr. 2002

Mission Need Statement (MNS), *Range of Nonlethal Force Capabilities (NLFC)*, 10 Oct. 1996

A Guide for Effective Peacetime Employment of Reserve Components Units and Individuals, OASD RA (M&P), Dec. 2000

Defense and Intelligence Abbreviations and Acronyms, Nov. 1997

Emergency Management Assistance Compact—Guide Book and Operating Procedures, National Emergency Management Association, Aug. 2000

Inside the Navy, Marine Corps Nonlethal Weapons Experiment Underway at Camp Pendleton, CA, Vol. 9, No. 29, Jul. 22, 1996

Military Support To Civil Authorities—Interagency Disaster Resource Guide, National Interagency Civil-Military Institute, Jan. 2001

National Guard Bureau Military Support to Civil Authorities Lessons Learned, 6 Sep. 2000

The Robert T. Stafford Disaster Relief and Emergency Assistance Act, as amended 42 USC 1521, Jul.2001

Joint Universal Lessons Learned System, Operation UNITED SHIELD

Nonlethal Weapons Joint Mission Area Analysis/Joint Mission Needs Analysis, Dec. 2000

Interservice Nonlethal Individual Weapons Instructor Course (INIWIC) Program of Instruction, Oct.2000

Presidential Decision Directive PDD-39 U.S. Policy on Counterterrorism 21 Jun.1995

PDD-62 Combating Terrorism—Fact Sheet, 22 May 1998

The Federal Response Plan Notice of Change 7, Feb.1997; FEMA, Number 229 Change 11

Federal Response Plan, Apr 1999

GLOSSARY

PART I – ABBREVIATIONS AND ACRONYMS

A

ACOG	advanced combat optical gunsight
ADS	aerial delivery system
AFDC	Air Force Doctrine Center
AFI	Air Force instruction
AFMAN	Air Force manual
ALSA	Air Land Sea Application
AMC	Army Materiel Command
AO	area of operation
AR	Army regulation
ATSC	Army Training Support Center

B

bldg	building
BLT	battalion landing team
bn	battalion

C

C4ISR	command, control, communications, computers, intelligence, surveillance, and reconnaissance
CA	civil affairs
cal	caliber
CANS	compressed air net system
CDC	crowd disposal cartridge
CMC	Commandant of the Marine Corps
COB	close of business
COC	combat operations center
COMDTPUB	Commandant, Coast Guard Publication
COMDTINST	Commandant, Coast Guard Instruction
CONUS	continental United States
COTS	commercial off-the-shelf
CP	command post
CQB	close quarters battle
CS	O-Chlorobenzyl-Malononitrile (riot control agent (tear gas))
CSSOC	combat service support operations center
cu ft	cubic feet

D

DA	Department of the Army
dB	decibel
DECON	decontamination
demo	demonstration
DM	designated marksman
DOD	Department of Defense
DODD	Department of Defense Directive
DODI	Department of Defense Instruction
DODIC	Department of Defense Identification Code
DODAC	Department of Defense Ammunition Code
DOT	Department of Transportation
DP	downed person
DPB	deployable pursuit boat
DTD	dated

E

EADBH	Eagle Duty Baton Holder
EEZ	exclusive economic zone
EOD	explosive ordnance disposal

F

FAST	Fleet Antiterrorism Security Team
FDP	food distribution point
FLEX	forearm/ankle cuff
FM	field manual
four M's	mind, mass, movement, muscle
FSE	fire support element
ft	feet

G

G1	Army or Marine Corps component manpower or personnel staff officer (Army division or higher staff, Marine Corps brigade or higher staff)
G2	Army or Marine Corps component intelligence staff officer (Army division or higher staff, Marine Corps brigade or higher staff)
G3	Army or Marine Corps component operations staff officer (Army division or higher staff, Marine Corps brigade or higher staff)

G4	Army or Marine Corps component logistics staff officer (Army division or higher staff, Marine Corps brigade or higher staff)
ga	gauge
GOTS	government off-the-shelf

H

HMMWV	high mobility multipurpose wheeled vehicle

I

IADS	Integrated Air Defense System
IAW	in accordance with
ID	identify/identification
IFF	identification, friend or foe
in.	inch
INIWIC	Interservice Nonlethal Individual Weapons Instructor Course
IPB	intelligence preparation of the battlespace
ITRO	Interservice Training Organization

J

JMAA	Joint Mission Area Analysis
JTTP	joint tactics, techniques, and procedures

K

KFOR	Kosovo forces

L

LBE	load-bearing equipment
LOAC	Law of Armed Conflict
LP	listening post
LVOSS	light visual optical screening system

M

M16	M16A2 rifle
M203	M203 Grenade Launcher
MARCORSYCOM	Marine Corps System Command
MCCDC	Marine Corps Combat Development Command
MCCM-GE	modified crowd control munition—ground emplaced
MCO	Marine Corps Order
MCPDS	Marine Corps Doctrinal Publication System
MCRP	Marine Corps reference publication

MCSF	Marine Corps security force
MCWP	Marine Corps warfighting publication
MDST	Monadnock Defensive Tactics System
MEDEVAC	medical evacuation
MEF	Marine expeditionary force
METT-T	mission, enemy, terrain and weather, troops and support available – time available
METT-TC	mission, enemy, terrain and weather, troops and support available – time available and civil considerations
MEU	Marine expeditionary unit
MIO	Maritime interception operation
MLEM	Maritime Law Enforcement Manual
MILSTRIP	military
mm	millimeter
MNS	mission need statement
MOAB	management of aggressive behavior
MOOTW	military operations other than war
MOS	military occupational specialty
MOUT	military operations in urban terrain
MP	military police
mph	miles per hour
MRE	meal, ready to eat
MTTP	multi-Service tactics, techniques, and procedures

N

NATO	North Atlantic Treaty Organization
NAVSOP	Navy standing operating procedures
NDC	Naval Doctrine Command
NLFC	nonlethal force capabilities
NLW	nonlethal weapon
NSN	national stock number
NSWC	Naval Surface Warfare Command
NWDC	Naval Warfare Development Command
NWP	Naval Warfare Publication

O

OC	Oleoresin Capsicum
OCAT	Oleoresin Capsicum aerosol training
OCONUS	outside the continental United States
OD	olive drab
OP	observation post

OPFOR	opposing force

P

PA	public affairs
PAO	public affairs officer
PASGT	personnel armor system, ground troops
POI	program of instruction
PSYOP	psychological operations
PT	physical training
PVAB	portable vehicle arresting barrier
PVC	polyvinyl chloride

Q

QRF	quick reaction force

R

R&D	research and development
RCA	riot control agent
RGES	running gear entanglement system
ROE	rules of engagement
RRF	rapid reaction force

S

S1	battalion or brigade manpower or personnel staff officer (Army; Marine Corps battalion or regiment)
S2	battalion or brigade intelligence staff officer (Army; Marine Corps battalion or regiment)
S3	battalion or brigade operations staff officer (Army; Marine Corps battalion or regiment)
S4	battalion or brigade logistics staff officer (Army; Marine Corps battalion or regiment)
SASO	stability and support operations
SB	shield bearer
sec	second
SECNAVINST	Secretary of the Navy instruction
SG	shotgun
SJA	staff judge advocate
SNCO	staff noncommissioned officer
SOP	standard operating procedure
SOTG	Special Operations Training Group
SROE	standing rules of engagement
SSC	small-scale contingency

T

TC	training circular
TF	task force
T/O	table of organization
TOC	tactical operations center
TRADOC	United States Army Training and Doctrine Command
TTP	tactics, techniques, and procedures

U

UN	United Nations
US	United States
USA	United States Army
USAF	United States Air Force
USAMPS	United States Army Military Police School
USCG	United States Coast Guard
USMC	United States Marine Corps
USN	United States Navy
USS	United States Ship

V

VCR	videocassette recorder
VLNL	vehicle launched nonlethal grenade
VMNLGL	vehicle mounted nonlethal grenade launcher

W

w/	with
WMD	weapons of mass destruction

PART II – TERMS AND DEFINITIONS

Air Force specialty code (AFSC). Primary job of a U.S. Air Force officer or enlisted person. Analogous to the Army and Marine Corps military occupational skill (MOS).

area of operation (AO). An operational area defined by the joint force commander for land and naval forces. Areas of operation do not typically encompass the entire operational area of the joint force commander, but should be large enough for component commanders to accomplish their missions and protect their forces. (JP 1-02)

battalion landing team (BLT). In an amphibious operation, an infantry battalion normally reinforced by necessary combat and service elements; the basic unit for planning an assault landing. (JP 1-02)

civil affairs. Designate active and Reserve Component forces and units organized, trained, and equipped specifically to conduct civil affairs activities and to support civil-military operations. (JP 1-02)

civil-military operations. The activities of a commander that establish, maintain, influence, or exploit relations between military forces, governmental and nongovernmental civilian organizations and authorities, and the civilian populace in a friendly, neutral, or hostile operational area in order to facilitate military operations, to consolidate and achieve operational U.S. objectives. Civil-military operations may include performance by military forces of activities and functions normally the responsibility of the local, regional, or national government. These activities may occur prior to, during, or subsequent to other military actions. They may also occur, if directed, in the absence of other military operations. Civil-military operations may be performed by designated civil affairs, by other military forces, or by a combination of civil affairs and other forces. (JP 1-02)

combat service support. The essential capabilities, functions, activities, and tasks necessary to sustain all elements of operating forces in theater at all levels of war. Within the national and theater logistics systems, it includes, but is not limited to, that support rendered by service forces in ensuring the aspects of supply, maintenance, transportation, health services, and other services required by aviation and ground combat troops to permit those units to accomplish their missions in combat. Combat service support encompasses those activities at all levels of war that produce sustainment to all operating forces on the battlefield. (JP 1-02)

command and control. The exercise of authority and direction by a properly designated commander over assigned and attached forces in the accomplishment of the mission. Command and control functions are performed through an arrangement of personnel, equipment, communications, facilities, and procedures employed by a commander in planning, directing, coordinating, and controlling forces and operations in the accomplishment of the mission. (JP 1-02)

commercial off the shelf (COTS). Describes items procured for military use through civilian contractors.

core capability. The set of functions that define a significant, stand-alone, operationally effective and suitable military capability such that, should no further development occur, the user will have received a significant capability.

designated marksman (DM). A rifleman stationed in an overwatch position above friendly forces, armed with a standard infantry weapon appropriate to his service that is mounted with a high-powered scope and loaded with lethal ammunition. The designated marksman uses his vantage point to identify agitators to forces below, eliminate lethal threats, and monitor the battlefield.

dispersion. A scattered pastern of hits around the mean point of impact of bombs and projectiles dropped or fired under identical conditions. In chemical and biological operations, the dissemination of agents in liquid or aerosol form. (JP 1-02)

fleet Marine force (FMF). A balanced force of combined arms comprising land, air, and service elements of the U.S. Marine Corps. A fleet Marine force is an integral part of a U.S. Fleet and has the status of a type command. (JP 1-02)

force protection. Security program designed to protect service members, civilian employees, family members, facilities, and equipment, in all locations and situations, accomplished through planned and integrated application of combating terrorism, physical security, operations security, and personal protective services, and supported by intelligence, counterintelligence, and other security programs. (JP 1-02)

government off-the-shelf (GOTS). Analogous to COTS, but describing items already procured by the U.S. government and ready for issue to military units.

host nation. A nation that receives the forces and/or supplies of allied nations, coalition partners, and/or NATO organizations to be located on, to operate in, or to transit through its territory. (JP 1-02)

intelligence preparation of the battlespace (IPB). An analytical methodology employed to reduce uncertainties concerning the enemy, environment, and terrain for all types of operations. Intelligence preparation of the battlespace builds an extensive database for each potential area in which a unit may be required to operate. The database is then analyzed in detail to determine the impact of the enemy, environment, and terrain on operations, and present it in graphic form. Intelligence preparation of the battlespace is a continuing process. Also called IPB. (JP 1-02)

interrogator-translator team (ITT). Marine Corps unit, usually working as a small detachment to a larger infantry unit, highly trained in interrogation techniques and fluent in adversary native language.

joint force commander. A general term applied to a combatant commander, subunified commander, or joint task force commander authorized to exercise combatant command (command authority) or operational control over a joint force. (JP 1-02)

logistics. The science of planning and carrying out the movement and maintenance of forces. In its most comprehensive sense, those aspects of military operations which deal with:

a. Design and development, acquisition, storage, movement, distribution, maintenance, evacuation, and disposition of materiel.

b. Movement, evacuation, and hospitalization of personnel.

c. Acquisition of construction, maintenance, operation, and disposition of facilities.

d. Acquisition of furnishing of services. (JP 1-02)

low-intensity conflict. Political-military confrontation between contending states or groups below conventional war and above the routine, peaceful competition among states. It frequently involves protracted struggles of competing principles and ideologies. Low intensity conflict ranges from subversion to the use of armed force. It is waged by a combination of means employing political, economic, informational, and military instruments. Low-intensity conflicts are often localized, generally in the Third World, but contain regional and global security implications.

NOTE: Also known as small wars or "brushfire wars," low-intensity conflicts are rebellions, insurrections, and the like characterized by adversaries dedicated to a cause but lacking command and control structure and modern equipment.

Marine expeditionary force (MEF). The largest of the Marine air-ground task force (MAGTF) and the Marine Corps principal warfighting organization, particularly for larger crisis or contingencies. It is task-organized around a permanent command element and normally contains one or more Marine divisions, Marine aircraft wings, and Marine force service support groups. The Marine expeditionary force is capable of missions across the range of military operations, including amphibious assault and sustained operations ashore in any environment. It can operate from a sea base, a land base, or both. Also called MEF. See also aviation combat element; combat service support element; command element; Marine air-ground task force; Marine expeditionary force (forward); Marine expeditionary unit; special purpose Marine air-ground task force; task force. (JP 1-02)

maritime interception operation (MIO). Means by which an exclusion zone at sea is enforced. Involves the visit of merchant ships, by armed boarding parties if necessary, bound to, through, or out of a defined area, to examine each ship's papers and cargo and to search for evidence of contraband. The provisions must be established by a sanctioning body and must be applied to ships of all nationalities.

military operations other than war (MOOTW). Operations that encompass the use of military capabilities across the range of military operations short of war. These military actions can be applied to complement any combination of the other instruments of national power and occur before, during, and after war. (JP 1-02)

military occupational specialty (MOS). Primary job of military officer or enlisted personnel.

military operations in urban terrain (MOUT). All military actions planned and conducted on a topographical complex and its adjacent natural terrain where man-made construction is the dominant feature. It includes combat-in-cities, which is that portion of MOUT involving house-to-house and street-by-street fighting in towns and cities.

military police (MP). Marine Corps and Army police forces designated by specific MOS.

maritime pre-positioning force operation. A rapid deployment and assembly of a Marine expeditionary force in a secure area using a combination of strategic airlift and forward-deployed maritime pre-positioning ships. See also Marine expeditionary force; maritime pre-positioning ships (JP 1-02 and JP 4-01.6)

nonlethal weapons (NLW). Weapons that are explicitly designed and primarily employed so as to incapacitate personnel or materiel, while minimizing fatalities, permanent injury to personnel, and undesired damage to property and the environment.

a. Unlike conventional weapons that destroy their targets through blast, penetration, and fragmentation, nonlethal weapons employ means other than gross physical destruction to prevent the target from functioning.

b. Nonlethal weapons are intended to have one, or both, of the following characteristics:

(1) They have relatively reversible effects on personnel or materiel.

(2) They affect objects differently within their area of influence. (JP 1-02)

nonlethal weapon capability set. A set of systems containing weapon systems, munitions, and protective equipment. The components are explicitly designed and primarily employed to provide additional protection to friendly personnel and to incapacitate hostile personnel or material, while minimizing fatalities or permanent injury to intended targets and minimal damage to property and the environment. They are not required to have zero probability of producing fatalities or permanent injuries but are designed and employed in a manner that significantly reduces those probabilities over the traditional employment of currently fielded military weapon systems, munitions, and equipment.

nonlethal technology (N-LT). Synonymous with NLW, but also encompassing more advanced technology that is not yet available for use by military forces.

Oleoresin Capsicum (OC). Also known as "pepper spray", OC is made completely from organic materials and is FDA-approved for sale over-the-counter in the United States. Causes severe and immediate burning sensation to mucous membranes when sprayed into face, nose, and eyes.

psychological operations (PSYOP). Planned operations to convey selected information and indicators to foreign audiences to influence their emotions, motives, objective reasoning, and ultimately the behavior of foreign governments, organizations, groups, and individuals. The purpose of psychological operations is to induce or reinforce foreign attitudes and behavior favorable to the originator's objectives. (JP 1-02)

rules of engagement (ROE). Directives issued by competent military authority that delineate the circumstances and limitations under which U.S. forces will initiate and/or continue combat engagement with other forces encountered. See also law of war. (JP 1-02)

table of organization (T/O). Total manpower authorized for specific unit, organized by billet and rank. TO& E is a U.S. Army term which encompasses T/O and T/E.

training aids and devices. Any item developed or procured with the primary intent to assist in training and learning.

weapons of mass destruction (WMD). Weapons capable of a high order of destruction and/or of being used in such a manner as to destroy large numbers of people. Weapons of mass destruction can be high explosives or nuclear, biological, chemical, and radiological weapons, but exclude the means of transporting or propelling the weapon where such means is a separable and deliverable part of the weapon. (JP 1-02)

Index

Made in the USA
Monee, IL
04 November 2026

67387221R00085